High Country Woman

High Country Woman

My Life on Rees Valley Station

Iris Scott

WITH GERALDINE O'SULLIVAN BEERE

RANDOM HOUSE
NEW ZEALAND

To the Scott family
— past, present and future

Contents

Introduction

A Long History

The New Zealand falcon, out hunting, shrieks from the top of a beech tree, *kek-kek-kek*. Hikers on the Rees-Dart Track, loaded down by their packs, walk the trail along the valley floor that's marked by bright orange marker posts. Far above, a sou'wester whips snow off the Richardson Mountains and tears it into spindrift clouds. In the valley, a breeze eddies through the tussock. In the bush that skirts the valley, dappled light falls on lime-green moss; the beech trees gently drop their leaves to the forest floor, where robins dart and chitter. At the valley edge, the Rees River races by, pale with powdered schist rock.

I scan the horizon, my eye ranging over the cattle. In this landscape, under my care, is extraordinary biodiversity and complex ecosystems. It's *my* natural habitat, too, this wide-open space. It's big country, this farm of mine: 18,000 hectares (46,000 acres). On the station's eastern border loom the Richardson Mountains, our 'undrawn fenceline' with the neighbouring station, the Branches, which extends to the Shotover River near Arrowtown.

Above the Rees Gorge, all the flat land on the eastern side of the Rees River belongs to Rees Valley Station; all the flat land to the west belongs to Earnslaw Station. The Rees, formed by springs at the head of the valley, gradually expands as tributary creeks feed into it on its journey towards Glenorchy, where it flows into Lake Wakatipu through a wide delta of shingle and schist-sand and knots of willows. Thirty-five kilometres from this delta, at the head of the Rees Valley, is the remote southern wilderness of Mount Aspiring National Park, its brooding, deep-green beech-forested mountainsides and high snow-capped peaks forming the back boundary of the station.

Previous pages: The Head of the Lake veiwed from Bennetts Bluff. Mount Earnslaw is at right; A tramper follows marker poles through the Rees Valley. Part of the first day of the four-day Rees-Dart passes through Rees Valley Station.

This is a landscape carved by the last ice age, rounded out by glaciers. Scattered through the forest and valleys are boulders of schist, their flanks green and orange with lichen. Some are huge; like a giant's plaything, they have been tumbled by the moving ice and then abandoned where they now sit along the forest edges. Isolated moraines — some, like Mount Alfred, of very considerable size — are all part of the ice age's glacial remains.

If steep crags are one of the natural geological boundaries of the station, then the creeks between the spurs are its major stock boundaries. They run down from each mountain peak, creating a physical border for each of the blocks on the run. Fences? Not in this part of the world. When pastoral farming began here in the late 1860s, fences were built purely to create small enclosures to occasionally hold stock or to protect domestic vegetable gardens. But extensive farming blocks tended to be unfenced on most sides, and on Rees Valley this is still the case today. In any event, fences all too often can be carried away by snow, avalanche, wind and rain. No ordinary farm fence can withstand the punishment.

Rees Valley Station is divided into two blocks: The Farm, which is freehold, and The Run, which is pastoral lease. Each block inside these larger blocks has a name, dating from the early days of human endeavour here. Some, like Invincible or Reef, are reminders of the valley's relatively short-lived goldmining boom in the late 1880s. There are plenty of other echoes of an earlier era throughout the property. Arthurs Creek, for example, is named after Arthur Ford, one of the station's early boundary keepers — the men who lived lonely lives in remote huts in the far back country, looking after the stock and keeping a keen eye on weather patterns. As storms loomed, they set out with their dog teams to move the animals back to weather-protected areas. Men rather than

fences kept the livestock within the station boundaries.

I'm always so keenly aware of the long Pakeha history here. Rees Valley Station was a part of the pastoral run Number 346, granted in 1861 to Robert Campbell. Back then, the run stretched from Twenty Five Mile (or Simpson's) Creek, now on the road between Queenstown and Glenorchy, to Twelve Mile (or Ox Burn) Creek, and included Mount Alfred and all the flat land between the Rees and Dart rivers. The station then passed to John Butement, who held the lease from 1874 to 1887. Next came the Valpy family, who had also left their mark on the station and the district by the time they quit the Rees Valley run in 1895. The Mission Church that they established on the station now stands in Glenorchy, to where it was moved in 1950.

Next was James Dunery, who proved to be a great entrepreneur, not only taking on the lease and the freehold of Rees Valley Station but also running his own carting business, transporting goods to and from the goldmines up the valley. He worked such long hours that they called him 'night and day Dunery'; he'd work a shift at the mine, then hitch his team to a wagon and begin carting. His livery and horse stables became the first woolshed on Rees Valley.

Before deciding to let go of the property, Dunery had begun work on several plans, one of which was to establish a new site for the homestead. He moved eight huge fifteen-year-old Wellingtonia trees from the original homestead plot beside Precipice Creek to what is now the driveway to the current house. He also planted what has grown to be a vast holly hedge (more of which later). These days, at the homestead, you can't escape Dunery's legacy: the gigantic Wellingtonia trees tower over the driveway, and the holly hedge that runs along the drive

The driveway to the homestead is like a tunnel: Wellingtonia trees on one side and a giant holly hedge on the other.

and two sides of the garden is about 2.5 metres tall.

After Dunery came the Scotts, and we have been here — apart from an eleven-year gap — since 1905. One hundred and seven years with Rees Valley Station: that's quite a number.

Henry Scott — generally known as Harry — was involved in establishing the Glenorchy Library, where his great-granddaughter, Kate, is now the librarian.

CATALOGUE

OF

—BOOKS—

OF THE

GLENORCHY

PUBLIC LIBRARY.

OCTOBER, 1911.

CHAIRMAN:

ANDREW FRASER.

LIBRARIAN:

D. McKENZIE, SENR.

SECRETARY & TREASURER:

HENRY SCOTT.

1912.

PRINTED AT LAKE WAKATIP MAIL OFFICE, QUEENSTOWN.

Chapter one

Land Girl

THIS REMOTE SOUTHERN LANDSCAPE wasn't always my home and my life. I grew up in the north, in the Auckland suburb of Manurewa, and then at Drury, just south of Auckland. In my early teen years I had become horse-mad, and my parents moved from rural Manurewa to a small acreage — what would now be called a lifestyle block, I suppose — where there would be space to graze a horse. My family weren't country people but somehow I felt that the country was where I belonged.

Growing up, I had, of course, seen photographs of the huge, wild and beautiful places of the South Island. By the mid-1960s I had a chance to actually see them for myself. In 1965, I was one of five women accepted into the Massey University veterinary science degree, which had a requirement for work experience: by the end of the course we needed to have spent six weeks on a sheep farm, six weeks on a dairy farm, two weeks at a meat works, and two weeks at a diagnostic station, and to have completed several weeks' practical work in a vet clinic. It made sense; after all, New Zealand was — and still is — a country whose economy depended on agriculture. Learning about farming was an important part of a veterinarian's education.

In 1967, Avril Phillips, one of my four fellow female students, and I hatched a plan to do our work experience in the South Island. Avril had never been there either. She wrote to Brian Booth, the stock inspector in Queenstown, and asked whether he knew a farmer who would take either one or both of us. He talked to Doug and Jean Scott at Rees Valley Station, and leaned fairly heavily on them to take a vet student. It wasn't until they had agreed that he told them the student was a girl! In the end

---

Previous pages: The Rees Valley Station 'Farm' — the lowland part of the property — in winter. The Rees River runs in the distance.

I alone went south; Avril decided to follow her boyfriend to Taranaki and asked me if I wanted to take her placement.

I travelled as far as Christchurch with Anne Tindall, one of the other women students, whose hometown it was. Back then, that was an exciting journey. We took the train to Wellington, then sailed overnight on board the Wellington–Lyttelton interisland ferry. My father and the ship's master had been young seamen together and Dad had written a letter of introduction which I carried. 'Dear Meaty,' he wrote. 'My daughter is travelling on your hunk of scrap!' My father's letter gave us a great opening, and Captain Meatyard invited us to the bridge for the early morning docking.

My father's friend had an impressive name: Edward Kitchener Garnet Meatyard, shortened by his mates to Meaty. The ship of which he was captain, the *Wahine*, was not even close to a heap of scrap, however. It was new and shiny, having been launched in 1961. A year after we travelled on it, the ship — no longer under Meaty's command — did become scrap after its disastrous holing on Barrett's Reef in Wellington Harbour.

Once in Christchurch, Anne headed up the Waimakariri Valley to Grasmere Station and I caught the New Zealand Road Services bus (the government bus services then spanned the country) to Queenstown. It trundled across the Canterbury Plains towards Fairlie, on into the Mackenzie Country and across the Lindis Pass to Central Otago. That bus trip gave me my first glimpse of the dry, tussock-clad foothills of the Southern Alps. It excited me. I saw it as the start of a six-week adventure; I couldn't have known that it would be the beginning of a lifelong adventure.

Jean Scott met me at Queenstown in the family car, a Ford Custom (the Scotts were dedicated to Fords, but more about that later). As we drove along the gravel road from Queenstown to Glenorchy and on to Rees Valley Station she told me a bit of the history of the area, which used to be called the Head of the Lake, rather than Glenorchy, by the locals. Among other things, she told me that the residents were very proud of the road, which had only been open for five years and which had made a huge difference to their lives. These days it's sealed and marked, a thoroughly modern road which easily winds around the flanks of the hills and down to several little bays on the eastern edge of the lake along the way. But prior to the unsealed road there was just a bridle track; stock, freight and passengers had been transported up and down the lake by boat for around 100 years.

There is a point along the road, as it rounds the curve at Bennetts Bluff, where everyone slows to a stop and gasps at the beauty of the huge landscape. From this high point on the road you can look up Lake Wakatipu to brooding, dramatic mountain peaks, the mighty Mount Earnslaw at the centre of them all. A few more turns of the road reveal a view of Pigeon and Pig islands and the flat lands at the Head of the Lake and Glenorchy. Jean stopped the car and switched off the engine, and I got out to look at the view. The lake was like a sheet of glass. Far in the distance the mountains towered, austere and snow-clad. I was spellbound.

As we drove on, I found myself looking up at the hills and wondering what on earth the sheep could be eating. There didn't seem to be any grass, but there were masses of an odd, ferocious-looking bush which I later learned was called matagouri. The green rolling hills of the North Island with their huge flocks of sheep suddenly seemed far, far away. My life as a high-country land girl had begun!

In my trusty Land Rover, the beast of many burdens.

It was late spring, and lambing was still in progress when I arrived at Rees Valley Station. The station then had a flock of Romney sheep, kept for producing fat lambs, on the lower paddocks. I was asked to do the lambing beat of the Farm block, checking a couple of times a day to ensure the ewes and lambs were all right. There can be problems when ewes lamb — the lamb is born dead, or the birth can be difficult and the ewe can get into trouble and may die. Then the lamb either has to be found a ewe prepared to feed it or it has to be taken away for hand-feeding. No matter the trouble, those Romney ewes became my responsibility. (Not so many years later Rees Valley got rid of its Romneys when Doug had to do the lambing beat one season and got fed up with how intensive the work was compared with surveillance of easy-lambing merinos!)

Doug Scott, my tutor, was a wise and experienced stockman. Muscular, sturdy and enormously fit, he spoke with the soft burr that's common in the south. On the station he was called The Boss, and in the district he was well respected as a good farmer and an active community member. He'd been involved with almost every local group and effort, including the library and the hall committees and the Rabbit Board. Doug enjoyed a good 'discussion', and would often use the opposition's arguments against someone else in a later debate. He had a hot temper but was always quick to make amends with anyone he had roared at. He was well read, especially on geography, American military history and British history.

Doug was of the 'pack horse and wire strainer' generation of farmers and had little affinity for modern-day machinery and techniques. He preferred the old ways: a man and his dogs and his horse, moving stock in the hills with time-honoured skill. Doug taught me how to read a paddock. Yes, when it's lambing time you need to check for the problems noted above but you also need to look at the feed in the paddock. How many more days of it are there? Are any fences damaged? I began to understand how much more I would need to learn, to look for and attend to.

At a young age, lambs are tailed and get an ear-mark, indicating which station they belong to. The Rees Valley mark is a 'fork and forebit'. One of my early jobs was to help set up temporary yards and move the lambs and their mothers into them so this could be done. I held each lamb as I wasn't proficient at the marking and tailing.

It became very quickly evident to the Scotts that I was a keen horsewoman. They had several farm hacks so I was assigned one to ride and given the task of checking on the calves. There really couldn't have been a more perfect job. I was in one of my favourite places — on a horse — and getting to spend plenty of time outdoors in this remarkable big new environment.

Once I got out onto The Run I marvelled at the unusual geography of the property. The Rees River delineates the western boundary of the station, and a topo map clearly shows the ten spurs (Twenty Five Mile, Big Devil, Little Devil, Long Charlie, Vee, Broadspur, Bridges, Muddy, Reef or 'Invincible', and Twelve Mile) that run from the Richardson Mountains down to the river flats. Each of those spurs is separated by a creek that feeds into the Rees. The Scotts and those before them had made great use of that geography — the spurs form each of The Run's blocks and their existence drove the grazing cycles and stock management practices that were particular to Rees Valley Station.

But a station also runs on a well-kept homestead, and the longer I spent at Rees Valley the more I grew to admire Jean Scott. Born Jean Reid, the only child of Scots immigrants, she was a gentle, clever woman. Her father, James, was the manager of the scheelite mine and her mother, Agnes (or, more usually, Aggie) had been the Glenorchy postmistress before

Mt Edward
Dart Glacier
Cascade Saddle
Mt Ansted
Mt Tyndall
Headlong Peak
Mt Cunningham
Rees Saddle
Cleft Peak
Lochnagar
Mt Clarke
Centaur Peaks
Twenty Five Mile Saddle
Lapith Peak
Mt Ferguson
RICHARDSON
Bridges Saddle
Twenty Five Mile Spur
Twenty Five Mile Ck
Big Devil Spur
Big Devil Ck
Little Devil Spur
Long Charlie Spur
Arthurs Ck
Vee Spur
Broadspur
Mt Earnslaw/ Pikirakatahi
Hunter Strm
Rees River
Bridges Ck
Bridges Spur
Muddy Ck
Cone Hill
Turret Head
FORBES MOUNTAINS
Earnslaw Burn
PARADISE
Diamond Lake
Mt Alfred/Ari
Lake Sylvan
Route Burn

A bird's-eye view of the topography of Rees Valley Station, including the station's boundary.

OLIVINE RANGE
Arawhata River
Mt Aspiring
KA TIRITIRI O TE MOANA
SKIPPERS RANGE
Mt Albert
SOUTHERN ALPS
Lochnagar
Mt Earnslaw/Pikirakatahi
Hollyford River
RICHARDSON MOUNTAINS
Dart River/Te Awa Whakatipu
Rees River
Shotover River
MAIN DIVIDE OF THE SOUTHERN ALPS
Kinloch
Glenorchy
QUEENSTOWN
Frankton
Lake Wakatipu
LIVINGSTONE MOUNTAINS
THOMSON MOUNTAINS
EYRE MOUNTAINS
6
to INVERCARGILL

Rees Valley Station located in the wider Otago region.

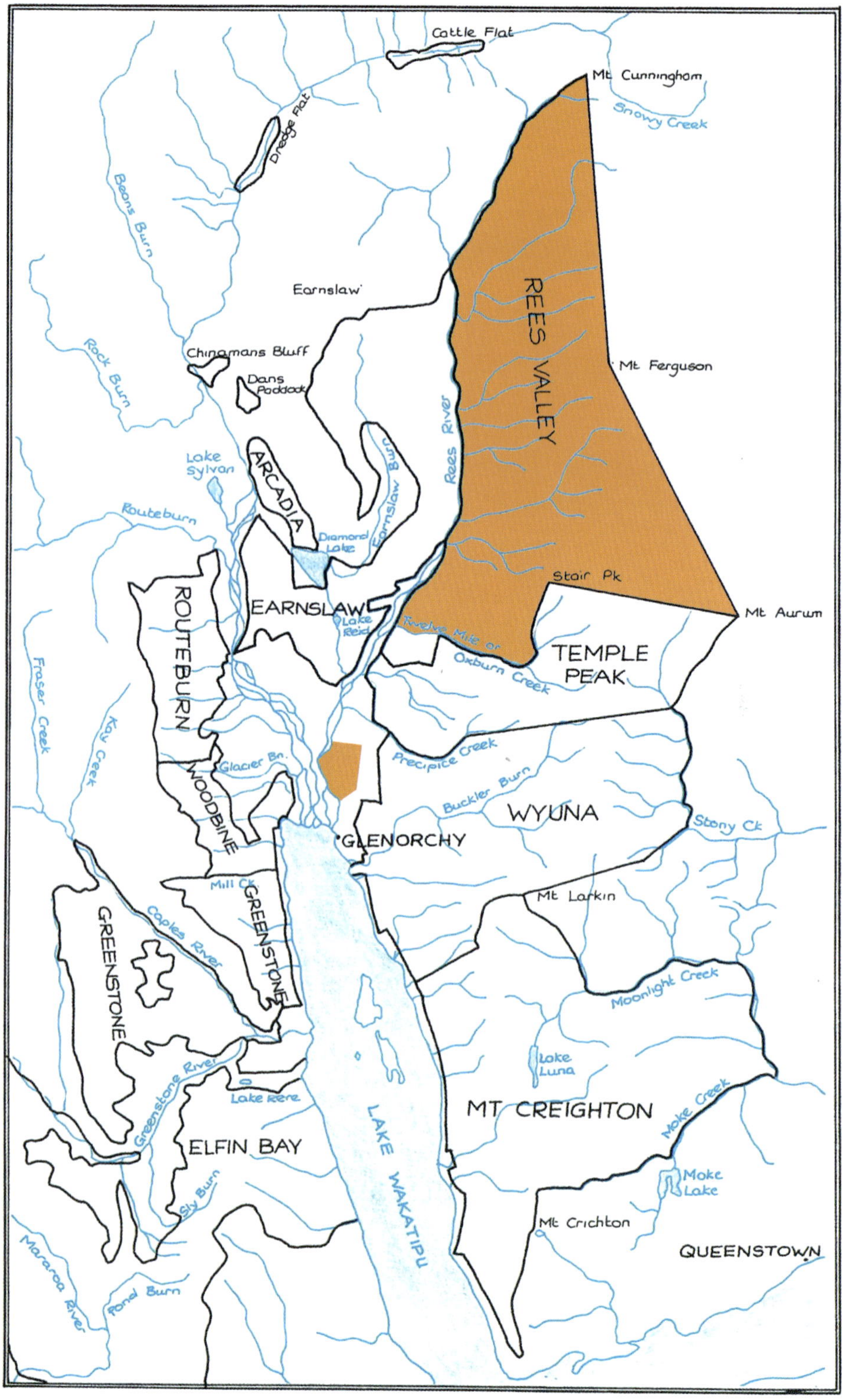

Cattle Flat
Mt Cunningham
Snowy Creek
Dredge Flat
Beans Burn
Earnslaw
REES VALLEY
Mt Ferguson
Rock Burn
Chinamans Bluff
Dans Paddock
Rees River
Lake Sylvan
ARCADIA
Earnslaw Burn
Routeburn
Diamond Lake
Stair Pk
Mt Aurum
ROUTEBURN
EARNSLAW
Lake Reid
Twelve Mile or Oxburn Creek
TEMPLE PEAK
Fraser Creek
Kay Creek
Glacier Bn.
Precipice Creek
WOODBINE
Buckler Burn
WYUNA
Stony Ck
GLENORCHY
Mill Ck.
GREENSTONE
Mt Larkin
GREENSTONE
Caples River
Moonlight Creek
Lake Luna
Greenstone River
Lake Rere
MT CREIGHTON
Moke Creek
ELFIN BAY
LAKE WAKATIPU
Moke Lake
Sly Burn
Mt Crichton
QUEENSTOWN
Mararoa River
Pond Burn

Above: Signpost in the Glenorchy township. Tourists have been coming this way for over 130 years. Opposite: The stations at the Head of the Lake in 1987. Rees Valley Station's two blocks, the freehold Farm and the leasehold Run, are in orange. Temple Peak Station was once part of Rees Valley Station but was carved off in the 1920s by Kate Scott as she struggled to make ends meet on the station after her husband, Henry, died.

her marriage. Jean had grown up in Glenorchy and had gone to boarding school in Gore before enrolling in the home science course at the University of Otago. At the end of her second year in Dunedin the Great Depression put paid to a university education: she was needed at home in Glenorchy, where she would meet and marry Doug in 1936.

There was almost an inevitability about that. In Glenorchy, which was, in that era, an extremely isolated place, the families of miners, small farmers and high-country run-holders were close-knit, and none more so than the Scotts and the Reids. After they retired, Jean's parents could often be found on Rees Valley, helping out with the work.

Jean had passed the Trinity School of Music letters exam, which enabled her to teach locals the piano. She was also extremely well read and one year she won Selwyn Toogood's *It's in the Bag* radio competition. The prize was to take Doug on a round-the-world trip. If Doug was called The Boss, then Jean was called The Chief or Bat, short for battle-axe — a name that seemed totally inappropriate given her gentle nature. Apparently the epithet had been bestowed one day when she'd shouted at the men returning to the homestead for their meal who had allowed their dogs to 'hoozle' the milking cow!

I was amazed at how busy Jean was every day. Although there was a separate cookhouse at the shearing shed for the shearing gangs, Rees Valley Station didn't have a standalone cookhouse for 'the boys' — the farmhands, working men, shepherds and musterers. In common with many Otago high-country stations, and in contrast, I'm told, to many Canterbury runs, the boys always came into the homestead to eat. Jean was very good in the kitchen and was known for her beautifully cooked dinners. No overdone grey cabbage for her. She was

Tools in the butchery.

adept at preparing a huge quantity of food, especially for the main meal of the day, served at midday, but she was not always on time. Jean was always relieved when the boys were running a bit late.

I appreciated the fact that Jean was enlightened and didn't expect me to do the inside chores just because I was a girl. In any event, at that time, a domestic worker lived in full time to help out, particularly with all the extra chores Jean had on her hands when the musterers, all needing to be fed hearty meals, were around several times during the year.

There was so much about station life that was new to me; the Scotts home-killed their own meat, for example, to supply both their own table and — at one time — those of neighbours. After all, there was no local butcher to pop down to for supplies. Rees Valley still has its own butchering shop, a small building with a chopping block and butcher's tools, including saws and cleavers and knives. On the wall there is a board once used for the neighbours' meat orders; the orders were written on tags and hung on their nail.

Beyond the back door of the kitchen, there was a shed divided into three areas, one of which was the food storeroom whose walls were lined with sheets of tin to keep out rodents. The other two areas were a dairy, where the cream and milk were separated and the butter churned, and the laundry with its copper and tubs.

At Arthurs Creek Hut, further up the Rees Valley and about two hours' ride from the homestead, there were big sorting yards for the stock round-up that the Scotts called the 'Fall Muster' (in autumn), and a lot of the station work was focused here; indeed, it was, at that time, the working centre of the Run. Boundary keepers, whose job it was to keep a weather-eye on the stock, were living at the Arthurs Creek huts at the time.

Above: A photo of sheep in the yards at Arthurs Creek, up the back of the station, in the late 1960s. Below: The old musterers' and boundary keepers' huts at Arthurs Creek today.

Staying out at Arthurs Creek — which the Scotts called 'the hub of the universe' — meant that the dogs and horses were fresh and ready to go for the day. Setting out from the homestead would, by contrast, require several hours of running the dogs and the horses before the stock work could commence. This is because Rees Valley Station has a slightly unusual set-up. The Farm (whose homestead is about 5 kilometres out of Glenorchy towards Paradise) and The Run are separate because of a quirk of geography that developed back in the 1920s when part of Rees Valley Station was divided off to create Temple Peak Station (more of which later).

As well as checking on the calves, several other farm tasks were allocated to me, including milking the house cow, Chocolate, and separating the cream. I very much enjoyed this task, and somewhat miss having house cows on the property these days. Temple Creek Station continues to have one.

One job that fell to me was peculiar to Rees Valley. Doug's father, Henry Scott, had had a dream to build a hydro-power scheme for the property. Henry purchased the disused water rights from a goldmine across the road, and water races, originally made to service the early twentieth century mining industry, were run through the farm paddocks of the station, bringing water for domestic and stock supply. The water flow offered the opportunity for more than just clean water. Henry was ahead of his time and didn't see his dream of hydro-electricity come to fruition, but Doug persevered with the idea and developed electricity from the water flow in the 1950s using a hydro-generator supplied by Turnbull & Jones in Dunedin and installed by Letch Sewell, whose family became intergenerational friends of the Scotts. The intake often filled with leaves, so if the lights dimmed during dinner it was my job to rush out, pull on my boots, and clear the debris from the inlet.

The plane in which I suppose I was wooed: Graeme and his Piper Super Cub in the late 1960s. The mountain behind him is the East Peak of Mount Earnslaw.

Jean and Doug Scott had two children, Graeme, born in 1939, and Heather, born in 1944. They'd both started out at the one-teacher Glenorchy Primary School, where the school day started late and went until 4pm because the school-bus timetable had to fit around the tourism traffic. The Glenorchy Motors bus would pick passengers up from the *Earnslaw*, which used to call at Glenorchy back then, and take them up to Paradise for afternoon tea. The children had to wait until the passengers had been returned to the boat before they were collected.

After the primers, they'd both endured long periods away from the family at boarding school in Oamaru. It was too difficult to get back to Rees Valley, except in the major holidays of May, August and Christmas; even the four-day Easter break wasn't long enough to warrant travelling (Graeme boarded at Waitaki Boys' High School, and Heather at Waitaki Girls') to Queenstown and up the lake by steamer.

By the time I arrived at the station, Heather was to be married to Gordon Collier, whom she'd met while she was a student at Otago University, and would soon move to Germany to live. Gordon had a scholarship to Mannheim University, and they would ultimately stay on in Germany for all his academic career, from which he has recently retired. (Both their sons and their daughter have come back to study at Otago.)

Graeme was 28, and seemed to me a rather shy country boy — you might say a late-starter. Given how isolated Glenorchy was and how small the local community, he hadn't had too many opportunities to meet women until a land girl landed on his doorstep. Many farming sons go away to get experience of working for someone else before they come home to the family farm. It's a way for them to get more farming education and knowledge and to see how another property works. So when

The Scotts in 1978. From left: Graeme, his mother Jean, his sister Heather, and his father Doug.

Graeme finished school he went to work on Matakanui Station, near Omakau. Going to another station, he could just be a worker rather than the boss's son, and when Doug lent him an old heading dog he began to enjoy it.

Graeme had flying genes, and had always had an ambition to fly. After he got his licence, he borrowed some money from his aunt and bought his own plane, a cherished Piper Super Cub that he'd land in the paddock near the homestead. Graeme and I sometimes worked together around the sheep and sheds, and eventually we started going for occasional flights in the Piper. We would get up at six in the morning, before the work day started, and fly over to the West Coast, Milford Sound and Martins Bay. After all, it was just over the hill! The flights were short, just there and back. On a clear morning it was spectacular. Below the snowline, dense rainforest cloaked the hills right down to the shore of the Tasman Sea, where endless lines of breakers were tossed onto the beaches.

At the end of the six weeks, Graeme took me to Queenstown airport to catch the plane back home to Auckland. At the departure gate, racked by nervousness, he ran his finger around the inside of his collar as he got up the courage to ask if he could write to me. I was a little taken aback. I had been perfectly happy thundering around on a horse and enjoying life in the hills without feeling any need to get romantically involved. It took him a while to convince me — four years of letter-writing in fact!

When Graeme asked me to marry him four years and all those many letters — and a few visits back down south — later I replied that unlike Jean, and Graeme's grandmother Kate before her, I wasn't any good at scone-making. He said someone else would make them! And fortunately for me, they almost always have.

In the cattle yards with the heifers.

Chapter two

Married Life

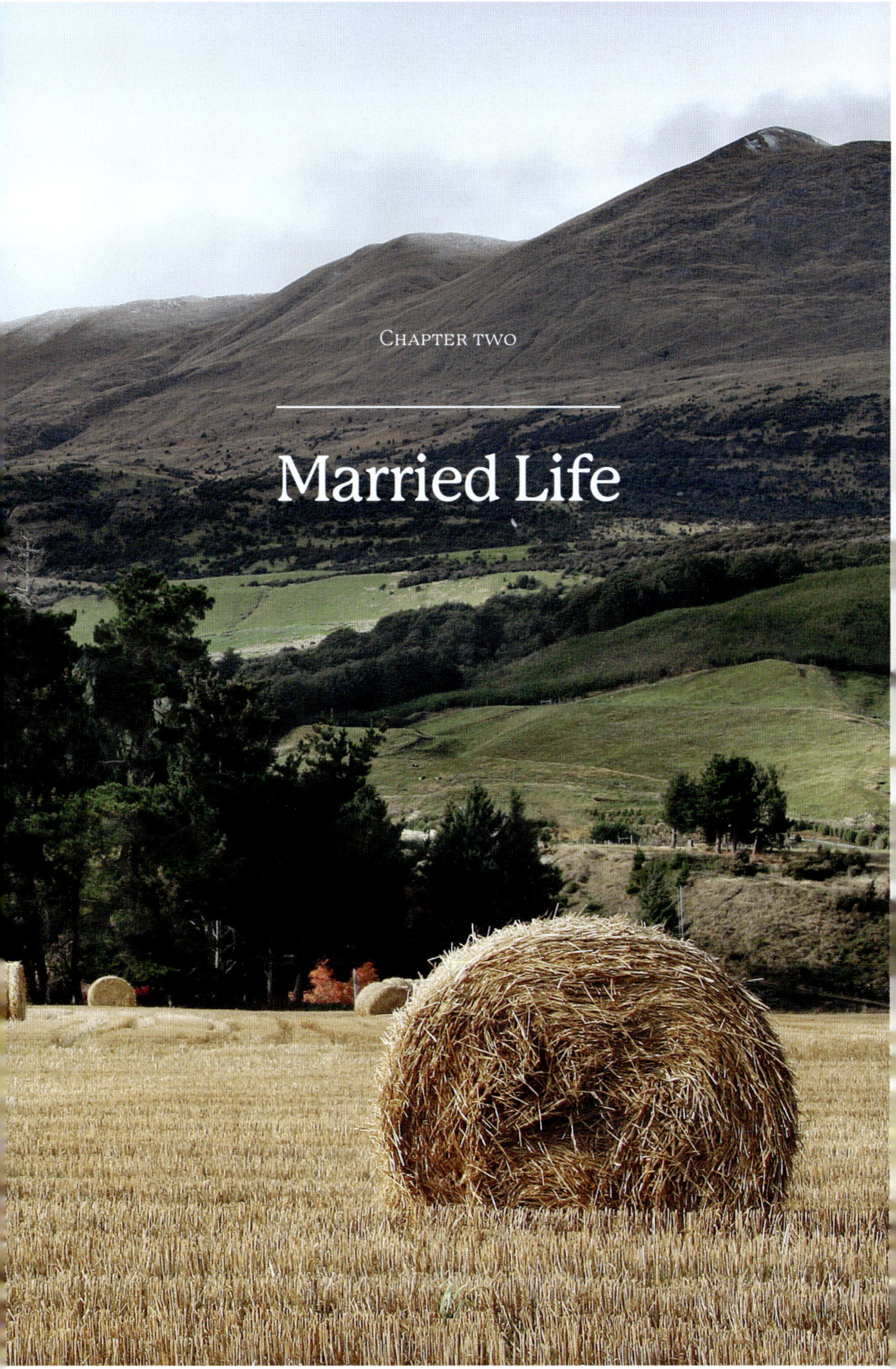

In February 1971 we were married, and with celebrations at both ends of the country— at my parents' in Drury and back in Glenorchy at the hotel — Graeme and I certainly felt we were well and truly wed.

When I first arrived, the station employed several people, but that began to change as farm incomes dropped from the late 1970s onwards. Unlike city life, there is no nine to five in farming; the work hours are dictated by the hours of daylight and the need to complete the job under way. The demands can be seven days a week as there are dogs to feed every day and always something else needing attention.

Initially, Doug was also out doing most of the stock work but eventually he had to retire from active service following heart trouble (he'd had a heart attack out on the hills), a worn-out hip and because of Jean's progressing Alzheimer's. He was devoted to caring for her.

Graeme and I decided that we would rather work the property ourselves so that any additional funds could be put back into the farm rather than go on wages. We then swapped houses. Graeme and I moved into the main house, built by James Dunery around 1904 and to which we added a top storey, and Doug and Jean moved into The Willows, a separate little house on the property.

Early on in our married life, we sorted out our preferences for responsibilities around the farm. Graeme enjoyed the machinery aspects of farming and I liked working outside with animals, so I gradually became more and more responsible for the stock work. Graeme had inherited his enjoyment of machinery from his maternal grandfather, James Archibald Reid, who had managed one of the area's scheelite mines. His grandfather was often at his workshop making things and, like many with a Scottish heritage, Graeme also liked the practical application of tools, machinery and engineering.

Previous pages: The Willows, my first home on Rees Valley Station. This page: Graeme and me 'raddling' ewes, some time in the 1980s.

And of course I also had my vet practice — which I'd set up using The Willows — to attend to. Even though there were six big stations and six small farms in the district, there wasn't a vet practice at the Head of the Lake when I arrived; the nearest vet was in Alexandra, 145 kilometres away. One did eventually set up in Queenstown a little later, but it is still a long way away from Glenorchy in an emergency. My rural vet clinic was originally a part-time service to the neighbours. After a while I moved the practice out of The Willows and into one of the station's sheds, before converting what had been the single men's quarters into a larger clinic.

Back then, it was unusual for women to go into veterinary studies. The Massey University School of Veterinary Sciences at Palmerston North had only recently opened as the first vet school in New Zealand, and no students had actually completed the five-year course or had graduated at the time I started. As an agricultural country we needed more vets, and demand for graduates was high. Before the vet school opened here, all New Zealand vets came from overseas, mostly from England and Scotland, and any New Zealander who wanted to study veterinary medicine had to go overseas, usually to Australia.

I was accepted as a student in the third intake. The lecturers were new and fresh, and so we were all young together as the school got established. The first year of university was an intermediate year of sciences, including the standard zoology, physics and chemistry as well as a maths certificate and a small introduction to botany. There were five women in the intermediate year, all hoping to get into veterinary school.

At the end of the intermediate year I went back home to Auckland for the summer holidays and began the anxious wait to hear whether my marks had been high enough and if I'd been successful. I had a job at the Tegel Chicken

Above: Graeme and me at a New Zealand Veterinary Association Sheep Society dinner in 1971. Below: Graeme and me in the late 1980s.

hatchery, which wasn't far from my parents' place and — very importantly — my horse. When I found out I had been accepted into veterinary school there was a celebration. Then to add to my great glee I also found that I could credit six weeks of practical from the chicken hatchery! At the time, the poultry industry was desperate for vets. I think they thought they'd found someone interested in chooks until I disabused them of that notion.

Many vets are employed in the New Zealand meat industry, and as vet students we had to complete two weeks' practical at a freezing works, covering meat inspection, food hygiene and animal welfare. I did my practical at a freezing works and meat-processing plant in Otahuhu in Auckland. To get there I would buzz along the Great South Road from Drury on my scooter. The scooter then went with me to Palmerston North on the Limited Express — cheap student transport.

I was very grateful to my sheep professor at vet school, Neil Bruere, for assisting me to get a job as a large-animal vet once I completed my studies. Everybody assumed that the women in the class were going into small-animal veterinary practice including cats and dogs — or, in my case, chooks — which was not what I had in mind at all.

Instead, I went to Tauranga to a dairy company job. The Kaimai Dairy Company then had several factories, including one at Katikati and one at Te Puke, which meant being rostered on to work at different locations. The general vet practice was owned and operated by the company which paid my salary, and this was then billed on to their farmer clients. That would soon change: the kiwifruit boom was just getting under way and dairy farms right across the district were being planted out

One of the several great walnut trees that Kate Scott planted, and which Doug Scott promised her he'd never cut down.

in kiwifruit. Eventually the dairy company almost shut down. However, by then I was setting off back to Rees Valley.

When I first returned, I repainted The Willows, and I worked a lot in the garden. Other women ahead of me, especially Catherine (Kate) Scott, Graeme's grandmother, had planted substantial orchards of fruit and nut trees there — peaches, plums, nectarines, apples, pears and quinces, walnuts and chestnuts — which still provide us with abundant supplies in season.

I continued to plant fruit and nut trees, younger replacement trees, thinking ahead to the next generation's food needs. I also kept up the large vegetable garden with long rows of summer crops. In summer I'd watch with chagrin as the shade from the large walnut trees that Kate had planted seventy years ago fell on the vegetable garden near the homestead. She'd made Doug promise not to cut them down — ever.

I learned to confront James Dunery's legacy, the huge hedge of holly he'd planted that was now over a hundred years old, vast and high and which took some effort to get under control. When it was time to trim the hedge I noticed helpful hands often failed to understand the subtleties of holly management!

And I learned to love swedes. Winter is too bleak and cold here for anything much to grow apart from swedes, which have been a winter crop as long as the Scott family has been on Rees Valley. The sweet variety was loved by both sheep and humans.

I loved the stock work, especially as doing it got me out into Rees Valley's wide-open spaces on a horse. Over time, though, I would come to appreciate how much faster a quad bike was — as well as being much easier to catch! — and how handy it is to have all the day's equipment in a tub on the front and my dog team running ahead or behind. When I had the children, either Doug or my mother, who had come to live on the station

Kate and Diane push Storm's puppies around in a toy pram.

in 1977, would bring a baby out to me so I could breastfeed and then carry on with my work.

I also came to value that great work beast the Land Rover; dog-carrier, occasional sheep transport, tools-carrier and fencepost-lugger. It wasn't long before ours looked like that of every high-country farmer: mud-splattered, with a thousand potentially useful small things in the cab, a dog or animal cage on the back, and a pile of hill-sticks alongside.

And then in 1992, after twenty-one years of marriage and with our family now including three children — Kate, then sixteen, Diane, fourteen, and Eric, nine — Graeme died of cancer.

He died at home after a very short illness. It was only six weeks between the diagnosis and his death. The family doctors and Deb Kennett, the local nurse, supported us brilliantly. Later, we took Graeme's ashes and scattered them near a mustering hut he'd built and that we had used for many years for our family holidays.

It was a tragic time. The previous year Jean had died, and then Doug had followed her within twelve months. Now I'd also lost my husband and workmate. It was devastating, and hard to fathom. Within twelve months, all those who had been so much part of the station had passed away.

I had a major decision to make. If I left the property, what might my children think in ten or twenty years' time about losing the land that their family had worked on for over a hundred years? With Graeme's death, I was struck by how it seemed as though history was repeating itself on Rees Valley Station.

I knew the Scott family story, of how distraught Doug had been when his widowed mother, Kate, had decided to leave the farm in 1921, when he was aged eleven. And how delighted he had been when they were able to return ten years later, when he was twenty-one and ready to take on the task of running the property.

In 1992 my children were too young to make an informed decision so I decided that I would stay until they were old enough to make a choice. The priority for me then became running the station. It was a daunting task, but with help from Margaret Blue, who, thank goodness, was a very capable farm manager (and also an expert at butchering), and Jeremy Close, whose family had a crib in Glenorchy and who came to work for me, I set to it.

Having to go out on the station and physically work hard each day on my own, without Graeme to talk through farm matters in the yards or over the kitchen table at lunchtime and in the evenings, didn't give me much time to grieve. I had stock to care for and my children to provide a future for. I was too busy to be lonely when Kate and Diane went back to boarding school in Alexandra after the funeral. Even though Margaret (who'd moved into The Willows) and Jeremy were close by, it was only Eric and me in the house to keep each other company at night.

I was keenly aware that those from the station's past had all gone, and of the impact of that. I thought how much more lonely it must have been for Heather, still in Germany, to have lost all her family in so short a time.

But I also knew that I had spent the best part of my working life, the last twenty years, working on Rees Valley. Plus I had my own intuition and observations to rely on; I had served my apprenticeship.

Chapter three

High-country Matriarch

Rees Valley Station has had a history of women at the helm. The word 'matriarch', advises the online Free Dictionary, can mean a highly respected woman who is a mother or it could be one who rules the roost and the family. It is a joke among my family to call me The Matriarch. They even put the title on a business card for me! As Doug was, I am also called The Boss (lovingly, though, most of the time).

I am not the only woman to have headed Rees Valley Station, although my daily tasks and responsibilities are altogether different from those of the first matriarch, Kate Scott, my late husband's grandmother. Kate ran Rees Valley Station many years ahead of me, and in a way that fully qualified her for the title of matriarch, too. She married Henry Scott in 1906, less than a year after he had made a successful bid for the lease of Rees Valley, taking over both the leasehold and freehold land from James Dunery.

Kate, née Cameron, had been dux of Otago Girls' High School in Dunedin in 1893 and was a schoolteacher at the time she met Henry. She had grown up in Alexandra, where her father had a hotel, and met Henry at the local Presbyterian church while he was working nearby at Olrig Station. Henry, generally called Harry, would no doubt have impressed her with his determination to succeed as a farmer. For a start, he was sensible. He'd already seen the terrible damage rabbits had done in Central Otago's light soils; he was looking for a property in a wetter climate, where rabbits wouldn't get out of control.

Henry was the son of Harry Scott, a farmer at Merton, north of Dunedin. Harry had been born in Roxburghshire, Scotland, and had sailed to New Zealand on board the *Storm Cloud* in

---

Previous pages: With Eric during the Hereford muster. We are on the Rees River flat near the Twenty Five Mile Creek mouth, with the Earnslaw massif behind.

1861. He had a family of twelve children, with Henry being number six in the line-up. Henry was known as a hard worker and a skilled musterer and stockman, 'a good man with a dog'. His large number of siblings came in handy when extra hands were needed at Rees Valley.

Kate came to the station as a new bride at a time when there were numerous staff to keep the whole enterprise running. An annual record to 31 March 1912, for example, notes the wages of staff including a head shepherd, under-shepherd, stationhand, ploughman and domestic and casual staff including shearers, fencers and odd-jobs people. Managing and feeding so many staff was a considerable task.

By all accounts Henry and Kate ran the station extremely well. Kate's farm-diary entries talk of jam-making and of planting fruit and vegetables; Henry's notes record buying bulls and the tasks various workers were undertaking, and where on The Run they were working. The diary also notes the annual routines and the special days, details of the cost of insurance for workers, and it even lists the shoes for the station hacks and draught horses. The diaries, which go back to 1905 and which we keep among our historical records and books in the lounge, are a time capsule of life at Rees Valley from those early days.

> Monday
> *Still warm. Kate, Jan and Mr Harris went to Paradise for a drive. Bull bought from Andersons.*
>
> Friday
> *Took shorn sheep up river. Turned bull out in Cash's with cows.*

These days they almost read like cryptic notes. But these simple words spell out much more as I read into them my knowledge

Kate and Henry Scott's wedding day in Alexandra in 1906. I always wonder why no one thought to remove the spade from the garden bed before taking the photo.

of the rhythm of work here. The Monday entry means that they'd have taken the trap out to the Dart Valley to visit the Aitkins at Paradise or the Reids at the elegant Arcadia, which were favoured outings on days off from station work. And the Friday entry about turning out the bull (which would have been sent from the stud farm to Rees Valley by steamer) and putting it out with the cows onto the Cash's block would have been an important day on the farm calendar, always occurring on a schedule, just as it does now.

It was life of toil and thrift and aspiration — and also of great achievement, as their hard work was rewarded with healthy stock and good feed crops. Two children came along, Doug and Jean, and there was a sense that they were building a future, that Scotts would continue to work hard and prosper here.

But then a health scare darkened their lives. When Henry was diagnosed with tuberculosis in 1915, he had to move out of the station homestead to protect the family, shifting to a special TB hut that had been provided for sufferers by the Health Department in an effort to control the spread of the disease. Henry's hut, called The Tent, still stands on the paddock near the shearers' quarters and woolshed. It's a small cabin with one door and one window, and the ceiling is lined with the canvas of the original tent roof. For a time, Henry and Kate went back to Central Otago, where it was hoped the drier air would be of benefit.

To no avail. By 1918, aged only forty-seven, Henry had been killed by TB. Kate had to step in and take over the station. She was initially assisted by her brother-in-law, Alexander Scott, with whom she eventually shared a section of the run which became Temple Peak Station. (This station is south of The Run on Rees Valley, and is no longer held by the Scott family.) But even with Alexander's help it all became too tough, and two years later she reluctantly decided to let go of the lease and move back to Alexandra, to the family home, Roseneath, to be close to her own kin and so that Doug and Jean, then aged twelve and ten, could finish their primary schooling and go on to secondary school. After finishing high school, Doug went to work for the Beatties near St Bathans, setting up his own team of dogs and getting back, with much delight, into the farming life that had been denied him by the shift to Alexandra.

What could not be foreseen was that ten years after the family had left, Kate and Doug were back at Rees Valley again. The Great Depression threatened. When it hit it would take all in its path into poverty and despair, from coastal city to the loftiest peaks of the high country. Few escaped the harsh reality of the Depression years — a time without money and jobs, and seemingly few glimmers of hope for the future.

At Rees Valley, the new leaseholders, the Kennetts, were struggling and hoped to do better goldmining than farming. They discussed it with Kate who tried again to sell the property, but there were no takers. Kate had little choice but to return to try and farm once more. Doug had pined for the property, so when the decision was made to return he could hardly wait. He had just celebrated his twenty-first birthday. He saddled up his horse, loaded his pack horse, got his dogs, and set out to ride south.

The day he left, the whole of the town of Alexandra was gathered in its finery in the main street, awaiting a royal visit by the Duke of York, who was to arrive by train from Dunedin. It was awkward, but there was no other way to get out of town and on the road towards Cromwell and Queenstown, save going down the main street, past the townsfolk. Photographs show Doug setting off astride his horse wearing a shirt and tie, his blanket-roll tied behind the saddle. The ride took him several days, his faithful dogs following him on foot all the way.

Doug, saddled up and with his pack horse and dogs, leaves Alexandra bound for Rees Valley Station after ten years away from the place he loved. It took him several days to get there.

Kate at the reins of a decorated trap in Glenorchy in 1935 to celebrate the jubilee of King George V and Queen Mary. The Rees Valley horses are George and Mary, who were broken in in 1911, at the time of the King's coronation.

Kate followed later by coach, and then took the steamer from Queenstown to Glenorchy. Jean, however, stayed on to work as a secretary in Alexandra. (She later married Clarry Rooney, the Ford agent. For many years after that the family only ever bought Fords. Family loyalty was uppermost.)

When Kate got back to the station in 1931, the staff consisted of a teamster for the draught horses, a cowman and gardener, a shepherd and under-shepherd, all living on the property. There was additional help on washing days, which were marathon all-day events in those times of boiling the copper. Kate was not out on The Run as we are today but rather managed the books, the homestead and the domestic staff and the garden. Doug assumed responsibilities for the farm work, his roles including head shepherd and farm manager. It was a major operation and uppermost in their minds must have been keeping money coming in to pay the wages. The staff were critical to the station's functioning — and no one wanted to see any of them join the growing ranks of the unemployed. (Later, as the Depression gave way to the war years, a shortage of labour sorely tested Kate and Doug's capacities.)

The Depression deepened, and farming had almost non-existent returns as exports fell and the price of wool nose-dived by 60 per cent. Yet the demands on the property continued, with the same yearly round of work and effort. Doug was to take no wages for ten years. It's hard to imagine the enormous challenges Kate must have faced, not just as a widow but also as a mother and provider, managing with Doug a substantial chunk of remote mountainous high country, with the station's sheep and cattle needing daily attention — all without any of the machines we have today.

There were few cars and trucks and no electricity. Everything was done by sheer hard labour. Kate's role, gleaned from her farm-diary notes, made first as a new widow and then ten years later following her return to Rees Valley, appears, like any woman's of the time, steeped in duties as a home-maker — cooking, baking, preserving, churning milk to make butter, gardening, sewing and mending, not to mention nursing. Not present are the words that might have expressed the grim reality of her situation. After all, she had twice given up the civilised existence of Alexandra to live in remote Glenorchy.

But she was dedicated to Rees Valley and determined to see the station prosper, and there couldn't have been a woman more suited to the challenge. And as a highly principled Christian woman, charity was her creed.

But Kate wasn't dour. Tall, strong and determined, she was a dashing person, and held the speed record for travelling from Glenorchy to Rees Valley by horse and gig! It was, no doubt, pulled by the station's two horses, George and Mary, who were broken in at the time of the coronation of King George V in 1911. The two horses were still in work twenty-four years later, in 1935, for the silver jubilee of King George and Queen Mary (George had actually come to the throne in 1910); Kate Scott drove the team to town for the celebrations. They were one of the major attractions of the festivities at Glenorchy, and a framed photograph of the horses and buggy, with Kate looking every inch the determined driver, hangs in the homestead. A poem was even written about them.

Scott's Horses at Glenorchy

Both were born in the same year and grew up staunch and sound,
And both were broke to harness when King George V was crowned.
One received the name of George named in honour of the King,
The other was Queen Mary for short she was called Queen.

And while King George in England reigned helped by his noble peers
These horses in New Zealand toiled for five and twenty years.
They ploughed in double harness and drew buggies on the road
And climbed those rugged mountains with many a heavy load.

It was joy and all the world when five and twenty years had flown
And King George and Queen Mary were still on England's throne
And far away New Zealand all their subjects did rejoice
And God Save The King was sung by many a cheerful voice.

And those two noble horses each with a royal name
To help the celebrations down to Glenorchy came.
Though they were old and weary, they held up their heads with pride
And harnessed to a buggy they trotted side by side.
And at the celebrations they proudly held their own
Because they were the namesakes of the sovereigns of the throne.

From *Poems* by Mrs Alice McKenzie,
author of *Pioneers of Martins Bay*

Out with the Land Rover near the mouth of Hunter Stream — a Rees tributary flowing out of Mount Aspiring National Park, the Leap Ridge (on Earnslaw Station) is in the background. Following pages: A family photo album montage showing life on the station in the late 1910s. The little girl is from a later era; she is Heather Scott, my sister-in-law.

DEFENDER

Chapter four

The Extended Family

When Doug married Jean Crichton Reid in 1936 it both typified and cemented the Scottish links that pervade Glenorchy — as I was to learn.

Jean's own family was typical of the first- and second-generation Scots migrants who'd peopled Otago and worked so hard to build up its agricultural economy. Her mother Agnes, who was originally from Stranraer in the south-west of Scotland, had sailed to New Zealand from London, where she'd trained as a seamstress. She had in fact intended to go to Australia but met fellow travellers on the ship who'd changed her mind about her destination. Not only that, but she changed her profession, too, when she was successful in landing the job at the post office. Aggie Reid — 'Wee Nana' as the Scotts called her; Kate was 'Big Nana' — was another formidable woman in the district, and we still laugh at some of her sayings. 'Over yonder in the field' always guarantees a laugh from the family.

James Archibald Reid, Jean's father, had had a Scottish education, practical and biased towards engineering. Invariably called Pop, he was always tinkering with tools, and when Graeme was young he was often at his side, helping out in the workshop and making things.

When Jean's parents were too old to look after themselves in Glenorchy, a former mine company cottage was moved on skids from the township to the station and set down across the paddock from the homestead. It was dubbed The Willows. Their move to Rees Valley was an example of how the station — in common with many other high-country runs — was home to several generations within one family. When James died, Aggie moved from The Willows into the main house to help Jean with domestic chores.

Previous pages: 'The Killers' — destined for the oven. Note the severely undershot jaw on the left-hand sheep — it has caused it to be culled.

At Rees Valley, Jean and Doug carried on the great working partnership exemplified by Henry and Kate before them, with the same clear division of labour, and also with two children, a boy and a girl — Graeme and Heather — to help and succeed them. From the start of her married life on Rees Valley, Jean had assisted her mother-in-law Kate with the domestic duties. It was considered her suitable role, and it was vital work.

The farm diary records this work, the days when Jean milked the family cows if the cowman was not available, and how social times were often based around games of cards:

> Saturday 30 May 1936
> *Mr J Reid, (Sen) Doug and Jean here for tea. And later young folk went to the dance. Mr J Reid and Mr Aly Reid spent evening here playing crib.*
>
> 5 Jan 1937
> *Picked gooseberries and made jelly. 1 and a half bottles of raspberry jam.*
>
> *Churned butter and baked. Made soap.*

In 1944, when she was in her late sixties, Kate left Rees Valley and returned to Alexandra to live with her daughter Jean. Clarry was still away at the war, and Jean's daughter Catherine was severely handicapped with cerebral palsy. Kate's help with her care was needed. She had saved Rees Valley as she had determined to do, and it was now in very good hands. (Kate died a few years later of cancer. It was my misfortune never to meet her.)

Doug had always liked the stock work, especially mustering and moving stock. If he could have been Marco Polo he would have been the happiest man alive. He was extremely fit and

~~24~~ 23 THURSDAY [236—129]
St. Bartholomew.

Weather ditto. Hedley ploughing self pottering about forenoon. Went to Head after dinner. brought up load of goods from Melri

Taylor paid his mutton &c with gold. paid him 36/ in cash

Cow away served with bull

~~24~~ 25 FRIDAY [237—128]

Weather ditto. Hedley ploughing self went to turnip sheep forenoon. Went to Head after dinner for load of trees. Wife and self went to school dance in evening. Killed sheep

~~25~~ 26 SATURDAY [238—127]

Breezy. Hedley grubbing turnips. Self doing odd jobs about forenoon. Pennies stumps after dinner took great pride in out-walking anybody and everybody. He just loved being out in the high country in the mountains with the sheep and his dogs. Indeed, he was accused of spending most of his life at Arthurs Creek, and his children used to joke about the strange man who rode his horse down the Rees once a week to get supplies and then disappeared in the early hours of the next morning. However, Doug needed to be up at Arthurs Creek, especially during the war years, when he was the only man working the stock. Jean probably wrote the farm diaries then, the entries marking his comings and goings:

Doug came down the Rees.

Doug stayed.

It must have been incredibly hard on Jean.

Doug had not been required to enlist in the Second World War because farmers' work was considered an essential industry. Such was the scarcity of labour during the war years that Doug once did the fall muster all on his own. It's hard to imagine just one man and his dogs out there mustering the whole property. Generally, it took twelve days for three or four men, each with a team of dogs, to complete it; Doug was out there on his own for several weeks.

And then it was our turn, Graeme's and mine, to become the third generation of Scott husbands and wives working in partnership. We were initially under Doug, an arrangement that gave me time to learn more about how extensive farming

---

A page from Henry Scott's farm diary from 1912. The economically written entries are fascinating: 'Taylor paid his mutton a/c with gold.' 'Wife and self went to school dance in evening.'

Graeme with one of his dogs, Roon, in the 1970s.

systems worked. High-country farming is a different method of farming, with low input and low production. It hadn't taken me long to appreciate how well this type of farming works in harmony with the land. You watch the rhythms of the seasons. Many of the farm animals are with us for years so we work carefully, always looking after them. It's symbiotic; if we look after the animals well they will look after us.

As Graeme and I took a more major role on the station, Doug and Jean began to step down from the active hands-on hard work, assuming the mantle of being there for the grandchildren. It was a great arrangement; there were three resident grandparents — Doug, Jean and, later, my mother, Marjory — to help out while Graeme and I were on the farm. The grandparents drove Kate, Diane, and later Eric, to school or met the school bus and looked after them while we carried on working.

My mother had moved to the station about six months after Kate was born. My father had died of cancer after a long illness — thankfully, not before he'd got to see his first grandchild — and she was finding life in Arrowtown, to where they had moved to be closer to us, a bit lonely. She lived in her own little house beside the homestead for twenty years. The children adored her. They would arrive home from school, and even if Graeme and I were at home they would run straight through the house and out to Granny, who had tea and chocolate biscuits waiting.

My mother Marjory was quite a character. She was upper middle class — her father had been the accountant at Harrods — and according to the strict social mores of England at that time, a young woman of her position was not expected to earn a weekly wage. It seemed inevitable that she would instead become a companion to an elderly aunt. That didn't appeal

My mother Marjory with me, Kate and Diane on our way to the Glenorchy Flower show in 1978.

to her at all. The only other 'respectable' jobs for someone of her class, as it was then considered, were either teaching or nursing, so, at age seventeen, she walked into a hospital in London and said to the matron, 'I want to be a nurse.'

'You are far too young. Come back when you're old enough,' was the matron's reply.

Shortly after that an influenza epidemic hit Britain. The hospital got in touch with her and said, 'Can you start straight away because we're so desperate. So many of our staff are ill and there are so many patients. Anyone who is still walking would be appreciated.'

Later she completed more nursing training, including in midwifery at Aldershot Hospital, a military barracks town not far from London. When my mother had her qualifications she saw a flyer on the hospital noticeboard calling for volunteers for the Foreign Nursing Service. She thought that sounded like an adventure and applied and was accepted. When she found out that the Foreign Nursing Service was sending her to Shanghai, she had to look in an atlas to find out where in the world it was! The Japanese invaded China and then Shanghai in 1937 and in Shanghai she saw some pretty horrific sights. Fortunately, she had already met a dashing ship's captain who had visited Shanghai and they had fallen in love.

My dad, Eric de Malmanche, was a Kiwi who'd been brought up in the timberlands of the King Country, in Raetihi, Mananui and National Park. De Malmanche wasn't, in fact, his real name. He was born Eric Vivian Kenny, one of seven children. When his mother died, the children were sent to various relatives. Eric was brought up by his aunt, Grace, whose husband was Robert de Malmanche, a descendant of early French settlers from Akaroa. They had no children, and Robert loved Eric and treated him like a son. He wanted someone to carry on the French surname so my father agreed to change his name, becoming

Eric Vivian de Malmanche. That's how as a girl I ended up with a French surname and a connection to Akaroa.

Eric left school aged twelve, having achieved his Certificate of Proficiency, and went to Auckland, to work initially for Subritzsky's on its coastal scows. Then, after he got his captain's qualifications, he went overseas and worked for the trading company Jardine Matheson in the South China Sea.

When my father first met my mother he was based in Singapore. They were married and had a long honeymoon, including a visit to New Zealand, which Marjory did not enjoy, saying to him, 'If you ever ask me to live in New Zealand I will leave you.' She had spent time in Gisborne, where the hills were still scarred by burnt trees, and the facilities seemed primitive to someone from Shanghai, 'the Paris of the East'.

On 15 February 1942, not long after their return from their honeymoon, Singapore fell to the Japanese, and my father was needed for combat in South East Asia. He left home one morning, saying to Marjory, 'If I don't come home to breakfast, you'll know I've been sent somewhere else.' That somewhere else was a secondment to the Indian Navy, which was under British command. He was to captain a minesweeper for the rest of the war.

My mother didn't talk much about all this unless you asked her, but fortunately she wrote it all down in a short memoir. She writes vividly yet calmly of almost unimaginable horror. The day my father left, things came to a crisis point. Rather than be lonely on her own in the house after a nightshift, my mother went back to the hospital in the middle of a bombing raid — 'By this time the Japs were keeping "Office Hours" with their raids, starting at about seven-thirty a.m. and knocking off about four-thirty p.m.,' she wrote — and had to leap into a drain, with two British soldiers, for safety.

Two days later the entire hospital – staff and patients — was evacuated onto the ship *Kuala*, which the Japanese proceeded to attack. 'I saw the gun crew rush out again. They were black and their clothing was in tatters. More crashing and lurching and when it quietened down again we came out on deck. At first I couldn't quite take it all in. The whole deck was scarlet. In the scuppers was a piece of canvas and sticking out of it was a pair of legs and feet clad in white duty stockings and shoes. "That little girl never knew what hit her," remarked a sailor.'

And that was just the start of three appalling years, as the Japanese rounded up the survivors of the ship and hundreds and hundreds of other men, women and children and forced them to march almost the whole way across Sumatra, from one filthy camp to another. Women died in their scores: of cerebral malaria, TB, diphtheria and of various forms of infection, in childbirth, of cancer, of a fatal form of dysentery, brain tumours, malnutrition and ulcers.

If there was a blessing it was that this was a company of women doctors and nurses of enormous grit, resilience and resourcefulness. My mother calmly recounts how Dr Marjorie Lyon — who emerges from the memoir as a woman of heroic stature, a great leader — operated on a woman with breast cancer without the benefit of anaesthetic; all she had was a spinal block. There was never enough food, and rats — and the occasional tiger — terrorised them at night.

My mother was released in 1945 and immediately began to search for my father. They were reunited in Singapore, where my father managed to get a shore job, becoming the shipping traffic supervisor for the Singapore Harbour Board.

When her camp was liberated she weighed only five stone and all her teeth were ruined. A dentist in Australia pulled all the remains out. When her health was sufficiently restored my mother started nursing again until they began hatching a family. I was born in Singapore in 1947 and my brother, Robert, in 1950. Still, life wasn't peaceful. The communist insurgency was making both Malaya and Singapore unstable and there was shooting in the streets. My parents thought they had 'been there, done that' in terms of war, guns and threats to their lives, and so it was that we came to New Zealand. This time round, Marjory couldn't get here, to peace and quiet, fast enough.

Our family life in New Zealand began first in Auckland, where we lived for a few months in Ponsonby with my father's aunt, who we called Grandma, while my parents built a house in Manurewa, then on the rural outskirts of Auckland. I went to primary school in Manurewa and later started at Diocesan High School in Epsom. Unfortunately, at that time, Diocesan did not teach the sciences that I wanted to study to those who were in the middle streams, and so I left after three years and went to Papakura High School, closer to home.

I was extremely happy to be able to have Marjory come to Rees Valley after my father died. She was a part of the extended family group here. By the time Graeme died she was well into her eighties, but still able to help me with household chores. Her being on the station was such an emotional comfort in the difficult times ahead. She spent the last eight years of her life at the Wakatipu Home for the Elderly, in Queenstown.

Firewood waiting to be fed into the wood burner.

Chapter five

Woman Alone

Many women run farm properties these days, but at the time I took over Rees Valley it was not so accepted. I'm sure that many people in the district assumed I would sell up when Graeme died, or that I'd put a manager on the property. It's not that long ago, but it was still the case in the early 1990s that at farm shows and tractor sales a woman farmer with money in her pocket, ready to buy a new tractor, would not be approached by one sales person. Looking back, it was pretty amazing that my decision to stay on and farm myself raised some eyebrows, but it was a fact that the actual hard physical work of farming and also the management was still considered a man's job.

Even so, I didn't encounter prejudice and I wasn't patronised for being a woman; I'd got past all that twenty years before, when I'd set up as a vet. (Well, most of the time. We still occasionally encounter sales people who ask to speak to 'Mr Scott'!)

For any farmer, man or woman, Rees Valley Station is challenging. Storms and snow can and do arrive unexpectedly, at almost any time of the year. Much of the land has to be covered on foot; the hills are generally far too steep for horses and there are only a few parts of The Run on the flat valley floor that are negotiable by vehicle. I need to be fit, as I still have to get up those steep sides, although these days I try to find younger and longer legs to do the real hill climbs if I can — or I wait until those legs are available for the job.

The Farm block is intensively farmed and cropped (with oats and swedes), just like any lowland farm. This part of the station looks as if it should have rich alluvial soils, but the soil is actually very poor; it's essentially ground schist, and it's not terribly fertile. It's only years and years of careful management

---

Previous pages: In the yards outside the woolshed in autumn. The mountain in the far distance, centre, is Mount Earnslaw's East Peak.

and nurturing that have made these flat hectares productive.

The wind is cruel here; it roars down from the mountains, as if that funnel only serves to increase the intensity of its bite. When I arrived at Rees Valley, there were pine shelter belts planted by Henry and later by Doug, but they ran east–west and it soon became clear to me that this was not much of a benefit to stock in winter; snow and ice would remain on the shady southern side. So I began a programme of planting coniferous shelter belts that ran north–south. It's very pleasing for me now, when I drive along the road from Glenorchy, before turning in at our gate, to look down from that elevation over The Farm, with its paddocks neatly tree-fringed.

The station has become my life's work and my life's adventure. It's been twenty years since taking on the challenge and accepting the responsibility of running it on my own, and both Kate and Eric are now old enough to be a real part of the station, working alongside me.

High-country farming is always a demanding yet rewarding life. Kate made the decision before going to university, where she studied English and art history, that she would return to Rees Valley to be a farmer. She now plays a large part in the community, and I think it benefits from her leadership and organisational skills, and her creative flair. She's active in the recently revived local branch of Rural Women — which seems fitting given that her grandmother was the founder of the branch, then part of the Women's Division of Federated Farmers, in 1931. Until recently she worked part-time as the office administrator at the Glenorchy School and still works part-time as the librarian at the Glenorchy Library.

She and I complement each other well as workmates. Kate, who knows her way around computer programs — something I know nothing about — is also responsible for the day-to-

day recording and management. We work with the stock as a partnership of mother and daughter.

On the days when Kate and I do the stock work our well-worn wardrobe matches the hard work. You'll not often find me out of my gumboots and work clothes and canvas hat, but Kate, despite being frequently in farm gear, also constantly surprises the locals with her fashionable attire and her collection of the latest shoes. She follows her great-aunt Jean Rooney in being fond of fashion. Doug's sister always managed to look like a fashion plate, even though it would have been difficult for her to get the latest clothes, living first here and then in Alexandra — both of these locations as remote as you can get from the fashion houses. Now, it is so different; Kate orders many garments and shoes online.

It's a bit of a mystery to me where Kate gets her love of domestic skills. She learnt to cook first at home and then when flatting. While she was studying at Otago University, Eric gave her a subscription to *Cuisine* magazine and she became even more inspired. Now we all benefit; the feasts at Rees Valley have become legendary.

Well beyond scones, Kate creates indulgent chocolate cakes, two-tiered sponges, huge casseroles and thick luscious soups; she even makes delicious elderberry wine, which brews and ferments for months before being bottled. Most visitors write on their Facebook pages and in emails about Kate's food talents and the delicious dishes they have enjoyed. In hindsight, I can see that I needn't have worried about my scone-making ability back when Graeme first proposed!

And thank goodness, because we often seem to have an army of people to feed. There isn't a fully stocked shop nearby

---

Vaccinating, with Kate, drenching, in the yards outside the woolshed.

so we have to be organised; Kate plans her menus ahead and makes sure we have plenty of supplies.

Still, it's all so much easier now than in the early days. Back then, like many high-country stations, Rees Valley placed huge bulk food orders twice a year: the shearing order, organised to arrive ahead of the gang, and the farm order. Those orders would include half a ton of sugar and whole chests of tea! All these food staples got sent up from Dunedin by Waddells, the food wholesalers, which specialised in country orders, and they'd arrive, consecutively, by train, steamer, horse and wagon or truck.

Those huge orders were still the norm into the early 1970s. But when the road to Queenstown opened, it all began to change. Now that the road is sealed between Glenorchy and Queenstown, nothing is impossible to get. Nevertheless it is a 100-kilometre round trip to the Queenstown supermarket so bulk-provisioning is still necessary. We buy big bags of potatoes, big bags of onions, and lots of loaves of bread, which we freeze down. We order dairy supplies such as boxes of butter and large cheese blocks, and a crate of milk and cream, which is delivered once a week. We also kill some of our own meat and still have the vege garden and fruit and nut trees.

Eric has also made the decision to work on Rees Valley, and is responsible for all our machinery and what you might call the high-tech part of station work. Being a helicopter pilot, he has changed the way we work with the stock (more on that later). He also works for Back Country Helicopters, a Makarora-based company, flying visitors to the Wilkin and other places. Eric assists Kate and me when we need a larger team, or when we're ploughing or harvesting. They are both great workers; Kate's an excellent tractor operator and Eric's the best fencer around.

My other daughter, Diane, is currently working in Western

Eric and I discuss the day's work. My veterinary clinic is in the background.

Australia, for a servicing company which does catering and cleaning at mining camps in the outback. She's also learning to fly fixed-wing aircraft there. Diane initially worked in hospitality and tourism in Glenorchy, often holding down — as is the way round here — several jobs at once that combined to be full time. She is a wonderful organiser.

All four of us are now the directors of the company we formed to own and operate Rees Valley Station. The family decision to stay on the farm has not led to an easy life. High-country historians accurately write of the reality of huge risks and of lives spent battling the elements. For all the 150 years that there have been high-country runs in New Zealand, working on these properties has always been a tough life of endeavour against mounting odds.

It's great to have this partnership with my children and to see them growing in their role as farmers. They do tend to defer to me as The Boss but I know that as time goes on I will be able to hand more and more of the decision-making over to them.

The Herefords grazing under matagouri.

Chapter six

Learning To Understand The Land

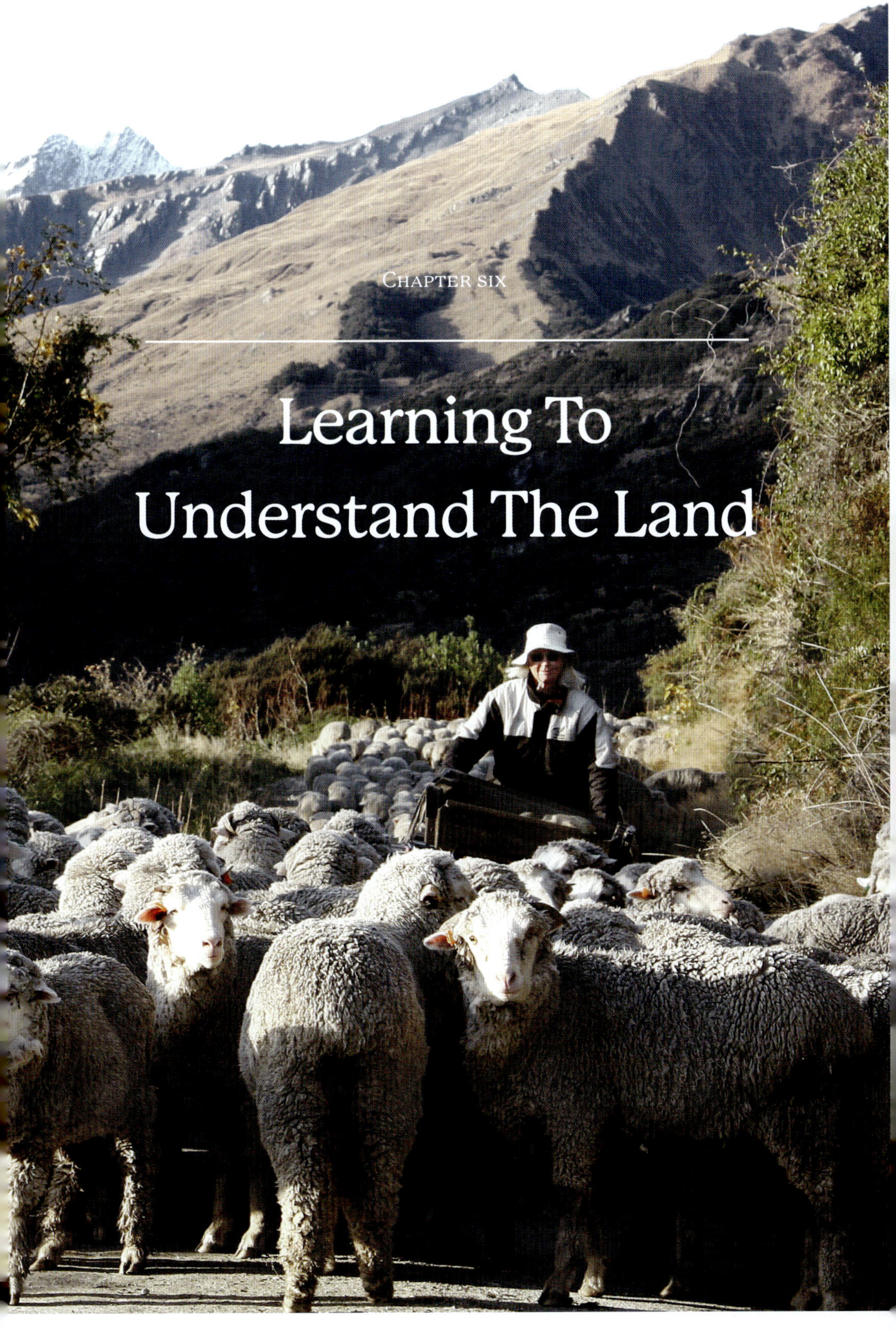

When farming leases were first offered to settlers by the government in the mid-nineteenth century, it was on the understanding they had a certain amount of time in which to stock their runs and build their homesteads. If they didn't achieve these objectives and couldn't get the stock there in time then somebody else was free to take up the lease. The government of the day really wanted the country developed for farming and set up incentives for that to happen. Leaseholders were not allowed to just sit on the land but had to do something constructive about farming it.

In the early days there were no set requirements for numbers of stock per run because nobody knew what the runs could carry. The more the merrier as far as the government was concerned because it added to the wealth of the nation. It was only after a period of time that it became obvious, in some cases, that the country wasn't able to carry the numbers. The amount of stock on some properties was pretty phenomenal. If the tussocks and other native plants became overgrazed, animal health suffered as a result of a lack of feed. Farmers converted native vegetation to 'improved pasture' and grew crops to maintain stock health. But when the plagues of rabbits arrived, the situation became grave.

The early high-country stories had a few myths laced into the truth, and their perpetuation has done considerable damage to the public perception of how the high country is managed. The famous and often cited Lady Barker — for example, widely quoted on the joy and thrill of huge tussock burn-offs — only lived in New Zealand for a few years, but her books about farming on a run in Canterbury became classics of high-country

---

Previous pages: Using the quad bike to drive the wethers from the woolshed to a winter block. I'd rather be on a horse, but I learned long ago that the quad bike is easier to catch!

station life. Classic doesn't mean authoritative, however! It's unfortunate that so much of her writing has passed into folklore and is taken as gospel. The truth is that high-country farmers have never been a reckless, devil-may-care lot.

Only a very few early high-country farming settlers in the Canterbury region were 'gentleman farmers', arriving in the new world to farm like lords of the countryside. Otago, in contrast, was settled by industrious Scottish immigrants, who brought their work ethic and diligence rather than huge wealth to the new land. Their goal was to work hard and earn enough to own their own farms. These were the people at the heart of the Otago high country. Their legacy and practical approach continues in the same families who have looked after and worked the land over four or more generations.

Even so, it is a fact those early English and Scottish high-country farmers were no more knowledgeable than somebody coming to farming from another profession today would be. They made mistakes. They didn't have the backup knowledge that has been acquired through some hard lessons. They didn't know then that if you burned off too much of the lower undergrowth you could destroy the soil or that if you brought in rabbits and let them loose to shoot at for fun, or just for something different on the menu, then you could destroy nearly everything growing in the high country. But they were not feckless; once they figured out how the land could best be cared for and the complex and often fragile ecosystems sustained, they did their very best to make intelligent land management decisions.

These farmers also didn't always factor in the weather, nor consider just how demanding hot dry summers and cold harsh winters could be. Snow is stunningly beautiful, but unlike skiers we high-country farmers do not have different words for snow. People who come to Queenstown and play in the snow have fancy words like 'powder snow' or 'spring corn'. Farmers just have 'wet and sticky', or 'powdery', depending on whether it can blow away or whether it freezes and turns to ice, or changes to rain. For us, snow can be dangerous; so much so that farmers tend to remember in detail the exact years of the big snows. They are memorable. My father-in-law Doug knew exactly which years had had big snows and could reel off the dates.

In winter only about a third of the station is not snow-covered. You never know when snow might fall, so it pays to be prepared. I check the weather daily on television and look days ahead to what might be coming. For serious weather conditions, we receive faxes from the Otago Regional Council which come through marked 'Urgent, Severe Weather Warning'. They are sent several times a day until the weather pattern moves away.

When really deep snow falls, our stock may not be able to get feed or shelter, depending on where they are. Of course we try to avoid such a situation. We plan for winter conditions by getting most of the animals off the tops and dark sides of the steep country before the cold weather arrives. However, weather can be unpredictable; snow can fall any time of the year. We may need to snow-rake tracks for the animals to walk out to areas where they can get feed and shelter. It's hard work, and in some areas the avalanche risk is high. More than once I've stepped out of a helicopter into chest-deep snow to get to the sheep, and more than once I have been grateful for the existence of these wonderful machines.

There was a time when New Zealanders were said to live off the sheep's back, on the export earnings of wool. In 1839 the

---

With Jerry and Rogue on a very cold winter's day near the homestead.

Ancare

first shipment of New Zealand wool was sent to Hobart, and the quality was noted. Stock agents in Australia began to get wealthy, as settlers moving to New Zealand needed sheep quickly. These agents bought sheep in Sydney for 2 shillings and 6 pence a head and then sold the sheep to New Zealand settlers for 25 shillings a head. For decades, the decks of New Zealand-bound trans-Tasman ships were crowded with sheep. Wool remained the big money-earner up to the 1880s. Sheep numbers increased, and the increased wool exports were bringing revenue that was critical for the new Dominion. Prices received a huge boost again when the Korean War began.

Those days have long gone. In the last two decades, in particular, there has been a huge decline in the prices sheep farmers receive for wool, yet the misconception lingers that high-country farmers are a wealthy farming group. Many New Zealanders see stock on large blocks of land — presenting an idyllic picture — and assume the operations to be hugely profitable. The reality is quite different. Economic returns are reliant on commodity prices, and these prices are determined by international impacts usually well outside our control. Careful financial management has been essential to ensure we survive.

When we had seventy million sheep, New Zealanders really did live off the sheep's back. Because of diminishing returns for farmers over the previous ten years, there has been a decline in sheep numbers, to only thirty-three million in 2010. Now, with an increased demand for wool, including merino, some commentators pick that the number of sheep is likely to increase.

When prices fell to painfully low levels, it became uneconomic to farm sheep for wool, and many farmers, especially those like us who concentrated on wool production, suffered. Through these difficult years we reduced the size of our sheep numbers so we could manage without needing to employ people and pay wages. Now we want to build up our own sheep numbers again and believe about 7000 sheep is about the right stocking capacity for Rees Valley.

But that will take time, and there have been years when farming here has been extremely challenging financially. Would it have been easier just to walk away? Possibly. Sometimes we have come close to breaking point, with many years where our bank balance would not shift out of the red. But since mid-2011 there has been good news for high-country farmers. It's been such a relief in the last years to finally have some positive news; the outlook is the most encouraging it has been in a decade. Interest rates have dropped, wool is fetching better prices. We are glad we have hung on.

When things are going well we do repairs and maintenance, making a start on the long list of work that costs money. Just like those before us, we have to wait for an improvement in returns and manage extremely carefully in the other years.

Even now, both Eric and Kate work off the station to earn their own income. Traditionally high-country farming is about making enough to look after the land and to educate the next generation, who will then become the stewards of the land. And, as I constantly remind myself, in so many other ways our life is rich beyond compare. We live and work in an outstanding environment. We work on land where we know who built each fence and who planted each tree. We take care in breeding our animals. We protect biodiversity and put pest programmes in place. Our major goal is caring for the land and its long-term future. We believe that high-country farming is the heart and soul of New Zealand's heritage. We love it. It is our life. We want to be here.

The days I spend on the station are marvellous to me, even after all these years. And I don't take any of it for granted.

Evidence of the ice age is everywhere you look: terraces, deep lakes and smooth glacial shapes high on the jagged crags. The terraces show that before the great glaciers began to retreat, Lake Wakatipu's edge once came higher up the lower slopes. You can clearly see this in the Rees Valley, a sharp horizontal line of terrace land stretching along the north-eastern side of the valley. These terraces are echoed by those of the Crown Range above Arrowtown, which run to where the lake once stretched, just as it extended to Athol in Southland, 300 metres above its current level, 18,000 years ago.

There is permanence, majesty and a reminder of the centuries past. The big spurs run down from the mountain peaks and almost vertical sides, an impenetrable steepness that defies even the most alpine of animal. Little can creep into these high sheer peaks so they form natural boundaries. It is almost impossible to try to master this landscape or make a mark on it. It is better to go with it, watch it, read it, and follow the lead of the land and the seasons.

A foggy winter's day on the Farm block.

559

Chapter seven

When More Gets Thrown At You Than Snow

THIS WEATHER HERE IS more than enough to make farming at the Head of the Lake challenging. But sometimes more gets thrown at you than snow, and you wonder where the money is going to come from to keep going. Some days it just all seems too hard. It would be easy to take the cash and run, as others have done before us, selling to international owners. Some run-holders have gone through a tenure review with the government, the result being that much of the tops of high-country properties have been absorbed into the vast Department of Conservation (DOC) estate, locked up as wilderness and never to be grazed again.

It's a flawed process, in my view, and just another example of the high country being misunderstood, of the history of government interventions made in ignorance. High-country stations are also of considerable interest to various lobby groups — conservationist, hunting, public access, for example — who historically have pressed the government hard on high-country issues. And a change of government can move the goalposts. It sometimes feels as though we have a change of landlord every three years and a new government may want to change the existing rules. For example, it's hard to believe that only twenty-five years ago the government of that time was encouraging intensification of farming in the high country. Now, this policy would be considered anathema by most, including today's government.

I have found myself, on several occasions, having to speak out against potential changes. As many government bodies are new to the needs of high-country land they often bring preconceived ideas to the table. It's somewhat frustrating!

It's fair to say that it has been extremely difficult for high-country run-holders over the years. Until 1914, for example, a

---

Previous pages: The entrance to the homestead after a light overnight snow.

tenant run-holder in New Zealand had to rebid for the lease from the state every few years, with every possibility of being outbid. Henry Scott, for example, purchased the lease of Rees Valley Station in 1905. Just two years later he had to bid for the lease again. It was no wonder that exploitation was common in those days; some leaseholders practised short-term thinking, getting whatever they could off the land before they lost the lease to a successor. Others, being optimists, put a lot of effort into improvements — because that's what defines farming as opposed to hunter-gathering — only to lose the lease. It was a disastrous formula.

It wasn't until 1948 that guaranteed tenure was introduced, providing security for the future. The system changed to a lease in perpetuity, with a right of renewal every thirty-three years and a rent review every eleven years. The welcome outcome of that was much more stable, sensitive and appropriate land management. The next decade, with its high wool prices and increased farm incomes, also brought land-management improvement opportunities to the high country. Rabbit boards were formed to get on top of the rabbit plague in an organised fashion, for example, and aerial top-dressing improved pastures.

Then, in the 1980s, a major reshuffle of government departments resulted in a split between the production management and conservation management of land. The New Zealand Forest Service, which had been responsible for 'good keen men' having huts and tracks and pretty free access to the back country for tramping and hunting, was divided into production forests (mainly pine plantations) and conservation (or native) forests. The newly formed Department of Conservation was no longer funded from the production forests; instead huts, tracks, weed and pest control became a taxpayer responsibility.

Similarly, the Lands and Survey Department was affected. Virtually all reserves came under DOC control unless required for Treaty settlements or the like. I remember the wry smile on then Prime Minister Geoffrey Palmer's face, captured by TV cameras, when he was proudly presented with the river valleys (Queen's Chain reserve land) that the green lobby had 'won' for DOC. He may have been aware of what a liability this gift was, the government now having to find the funds for weed and pest control especially, but also fencing.

Government departments were charged with operating profitably, and Treasury did not take long to point out that it cost more to administer pastoral leases than the rents returned. Raising rents to cover bureaucratic salaries was clearly not an option as farming was becoming less profitable — the increasing costs of farming were unable to be passed on by the primary producers, who could not operate as a 'cost-plus' economy. The government wanted to exit the business of being a landlord, and tenure review was suggested because pastoral leases do not include a right to freehold.

Then, in 1998, the Crown Pastoral Land Act was passed into legislation. It quickly proved to be a watershed change. Designed to address a raft of issues to do with Crown land, it also covered the areas of conservation and heritage, and included detailed regulations for pastoral leases and occupation licences, via a process known as tenure review.

Tenure review is voluntary, and anyone can apply to have their land assessed; to date, most of those run-holders who have gone for review have been looking to offer up a portion of their land to state control, mostly to DOC. Why? Economic pressures. Over the last ten years, during which farm income has been minimal — even loss-making for many high-country properties — opting for a tenure review and releasing part of the land has decreased the burden of lease-rental outgoings. The remaining land can then become freehold and sold, or split up for development. For cash-strapped farmers, this is

Caro and Kate on horseback during the autumn cattle muster.
The mountains behind are the Humboldt Range.

undeniably an attractive option. But now that the government is not in such a strong financial position, and is beginning to realise that more land under state control costs taxpayers more in upkeep, this process has slowed.

In general, land that could be farmed sustainably was offered for freeholding, while land of high conservation or recreation value was to be 'surrendered to the Crown'. For those properties where there was little land desirable to DOC, freeholding was unaffordable. For those where much of the property was at high altitude or had forests that the farmers wanted to retain, there was little incentive to enter into tenure review.

We watched the process as it dragged on for years, turning our farming friends grey and stressed. We made a resolution not to let this happen to us. We heard DOC proudly announcing that Molesworth Station (a property in 'full Crown ownership' for a very long time) would be managed for 'multiple purpose' — conservation, recreation and production. We thought, 'That's what we've been doing all along. Perhaps they would let us freehold and continue as we are!'

We volunteered for tenure review and duly were inspected by officials and their experts and NGOs, such as Forest & Bird and Fish & Game. The Rees Valley pastoral lease of 18,600 hectares is about a tenth of the size of Molesworth. We were offered 200 hectares freehold, 400 hectares freehold with a covenant that we could do nothing to it — even fences had to remain as they were — and the rest to be returned to the Crown. We asked what would become of it, curious to know why we had got our land use so wrong. The response suggested either they did not know or were not prepared to tell us!

We put in a counter-proposal, offering a modest boundary adjustment to make a more practical outline for Mount Aspiring National Park and proposing freeholding the lot, with QEII National Trust covenants protecting any conservation value the government identified. Our view was that the savings to the taxpayer of such a simple tenure review would far exceed the value of ongoing rents, and, because of the covenants, the nation would be spared worrying about us getting rich by flogging off bits to developers. Our proposal was declined.

So we are not fans of tenure review. Our preference at Rees Valley is to work with an organisation like the QEII National Trust, which has a fit-for-purpose covenant with landowners. Under the Trust, land is freehold and protected through the generations. The agreement is legally binding. The land cannot be sold off for development and there is permanent protection from damaging change. Members of the public can access the covenanted land with the landowner's prior permission.

The Trust is a unique and visionary New Zealand organisation, and covenanting land with its help is becoming more widespread. There are now over 3000 registered covenants covering more than 100,000 hectares, with many more covenants under way. As they accumulate there are increasing benefits to the environment. Land under covenant cannot be 'developed', that is, cut into small blocks with houses popping up on what was previously iconic landscapes and open-farmed country. We believe the long-term multiple-use system of the land under the Trust is better than the state taking full control and the burden of upkeep falling on taxpayers.

Owning and managing a high-country run in New Zealand tends rarely, if ever, to be financially rewarding, and is not a venture resulting in wealth and millions. Careful management is required to try to balance the books and at the same time to achieve sustainability of the agricultural business.

There's a lot of anxiety around the fact that many New Zealand high-country stations are being bought by new owners, often wealthy international people, some of whom have bought several stations near Queenstown and the head of Lake Wakatipu. These stations are no longer owner-operated and often have management teams running the properties. Some would argue that this is a major and regrettable change to the high-country tradition of farmer-operated properties.

But there are upsides, I believe. It's only 150 years since nothing was known about farming the high country of New Zealand. What has been learned in that time, that information, is now available to anyone starting out on a high-country property. So a new owner is much more likely to succeed, especially if they've got capital to do what needs to be done. Ironically, some high-country run-holders, who've been struggling for years, don't have the resources to do what they would like to do and what they know needs to be done. So in that respect it is good to see money coming into the high country from newcomers who have the resources.

Often overseas owners have an interest in the conservation value of the land, as we do. They make a contribution to looking after the property, through both conservation and farming. This is very positive. If a new owner buys a high-country property because they are committed to it, then that to me is a good thing. A good example is American record producer Mutt Lange, who bought Motatapu Station, at the back of Wanaka, with his then wife, the singer Shania Twain, in 2004. He has since bought Glencoe Station, on the Crown Range, and Coronet Peak Station, and together the three stations make a huge and contiguous block of land on which he is spending hundreds of thousands of dollars in plant and animal pest control and tree planting. That's to be saluted.

The Rees River, not far from the homestead. In the distance are the Twelve Mile Spur, Stair Peak and one of the ridges leading up to Temple Peak, part of the Richardson Range.

Since I arrived at Rees Valley in 1971 there have been changes in the ownership of both large and small blocks in the district. International owners have purchased three of the nearby stations, Wyuna, Woodbine and Closeburn; owner-operators are working Rees Valley, Earnslaw and Temple Peak; and, as a result of Treaty of Waitangi settlements, Ngai Tahu now owns the Greenstone, Elfin Bay and Routeburn stations.

What I find more worrying is real estate development. Farms are being cut up, the land divided into small plots for housing or lifestyle blocks. We appreciate that it is valuable real estate, especially if near a lake, but the values of conservation, recreation and productive farming are being lost to short-term rewards. In the meantime, the only farming that goes on is simply keeping the grass down until the developers are ready to complete their builds. They then move on and look for another block to develop. And as a greenie it bothers me to see the extremes of environmental management on these lifestyle blocks. It's great to have everything pin-neat, but you can't treat the high country as if it were a city section. We value having stinging nettles under our trees and around the sheepyards, for example, because they mean life for admiral butterflies.

Around the lake edges of Central Otago, this change is very noticeable. More and more areas have been cut up for housing and holiday-home sections. Only the current economic climate has slowed the process.

Old horseshoes in the tack shed. We currently have eighteen horses on the property. Following pages: Rogue and Jerry move a mob of merinos from the woolshed to the winter block on the Twelve Mile Spur after they've been crutched and eye-wigged.

Cheese Puffs

This is Jean Scott's recipe, and we think it probably was originally from her mother Aggie Reid.

1 large cup flour
1 cup coarsely grated cheese
1 egg, beaten in a little milk
2 teaspoons baking powder
enough milk to mix to a soft dough

Place spoonfuls on a cold greased floured tray and bake in a hot oven for 15 minutes.

Walnuts and quince from the homestead orchard.

CHAPTER EIGHT

Marvellous Merino

SPRING ARRIVES. IT SEEMS as though, all at once, everything is getting born or shorn. It's our busiest time of the year. Calves turn up, and then a little later the lambs do too. But it's the shearing that's the true culmination of a full year of animal husbandry. It's part of the science of taking care of sheep and it's a skilled operation.

When the shearing gang is in the shed there's a flurry of shearers, wool handlers, presser and wool classer, sheep, blades and bales. There's fleece on the floor, fleece on the classing table, fleece being sorted and packed into bales. Outside there's barking and whistling and little storms of dust as dogs move sheep to and from the yards. It's a time to admire the endurance and resilience — in fact, the pure grit — of merino sheep. They are the most remarkable animals, and have a fascinating history to match.

Merino became the treasured sheep breed of Spain in the 1700s. Spanish royalty had such a monopoly on the breed that it was a criminal offence to export any merino out of the country. Only a select few were ever sent away as gifts to 'special' friends of royalty, given as if they were precious metals or gems; or perhaps as bribery to garner favours.

High-country merino sheep often live independently for months on our steep mountain blocks, sometimes at below zero degrees. The fleece protects them against the conditions, from icy cold, freezing snow, pouring rain and biting wind. Growing fine and dense, the wool has exceptional natural properties.

Working closely with these animals, we realise they are extraordinary in other ways too. Far from the high street and the giddy heights of sporting events, where merino garments are

---

Previous pages: On The Farm on a foggy morning.

worn, this breed of sheep demonstrates an amazing stamina. A simple beast? We don't think so. In the high country, merino need to be acutely aware of their location and where they place each foot. They're out on narrow rocky outcrops, beside precipitously steep gullies, up almost vertical hillsides. They seem to have another sense, constantly aware of the position of others in the flock, and they also sense danger quickly.

Today's merino sheep in New Zealand are much bigger than their forebears from Spain. Back then a typical merino weighed only 22 kilograms or less and produced a very small fleece. Years of good animal husbandry have seen merino sheep become stronger and bigger, with a fleece of fine, dense fibres. They are now bred for good animal health, to suit the local environmental conditions and for the best economics.

The early merinos at Lake Wakatipu on William Gilbert Rees's North Run were imported from Australia and carried aboard sailing ships across the Tasman. It's hard to imagine sheep on the decks of a ship. To get to the new farms after offloading, they were driven along steep and perilous rock-strewn mountain tracks, often pushed by inexperienced drovers to cross cold mountain rivers and streams, and to trudge through dense scrublands. The drovers had scant knowledge of the area and maps were incomplete, so sometimes the flock had to be turned back and another track taken.

Just to get to the Head of the Lake from the Otago coast took days and days of effort. Accounts by Alfred Duncan, who was the drover of the first merino flock to walk out to the Head of the Lake, recall a journey along the lake's edge that took more than three days, travelling through land where there were no established trails. Merinos surely proved their hardiness and endurance.

Merinos have been on Rees Valley Station since the beginning of the run's pastoral days. In fact, our station is the furthest west of any New Zealand run stocked with merino.

Henry Scott's farm diaries, from 1905 on, record that the station had a carrying capacity of about 7000 sheep. My father-in-law, Doug, ran up to about 8500 sheep, but he only had a handful of cattle. As we have built up cattle numbers we've reduced sheep numbers to what we now think is the right balance: about 200 Hereford breeding cows and their replacements and currently about 4000 sheep, half of which are ewes and half of which are wethers.

We are proud of the merino wool we produce, especially because wool is a free-range and sustainable natural product. Not only do people now realise the benefits of wearing wool, especially merino with its exceptional qualities, but there is also an awareness of the downside of textile production using industrial petro-chemicals. Furthermore, it's recognised that in New Zealand sheep are farmed under strict ethical methods.

In 2008 Kate was invited to the annual Slow Food Terre Madre symposium in Italy, to attend sessions not about food but about naturally produced fibre. The Slow Food movement is well known. It was started as a rejection of the fast food industry and to staunch the loss of many traditional foods and their production. Now many more people are realising how important it is to grow and harvest traditional foods. In our own country there is more interest in cultivating vegetables and herbs and enjoyment of the growing process, which is then followed by the joy of harvesting your own fresh produce.

The organiser of the natural fibres section section of the Slow Food Terre Madre symposium was Elena Schneider, whose family company owns big woollen mills in Italy that buy a great

Merinos in the yards after blade shearing.

deal of New Zealand merino. She brought together natural-fibre producers from all over the world. It felt to Kate a bit like a UN meeting of fibre producers: French, Indian and Italian linen producers, South African mohair farmers, Mongolian, Chinese and Iranian cashmere producers, llama growers from Argentina and Bolivia, alpaca farmers from Peru and Argentina, silk growers from India and China, hemp producers from Italy, sisal producers from Brazil, jute producers from India and wool growers from Australia, Patagonia and South Africa — and Glenorchy! Held in the International Year of Natural Fibres, the drive behind the conference was the creation of an international network that would work in a way not unlike the Slow Food movement, promoting the benefits of natural fibres of all sorts. It feels good to be a part of a broader international movement such as this.

Wool is having a similar revival to locally produced food, with an appreciation of its sustainable natural production. There is a yearning for natural things, not just on the part of green crusaders but also fashion designers, carpet and textile designers and manufacturers. Dedicated international advocates, such as the Campaign for Wool, which has Prince Charles as its patron, are promoting wool throughout the world. Renewed enthusiasm is also being shown by New Zealand wool producers and designers. For years, the large international petro-chemical giants were able to throw huge dollars into marketing and increasing the use of cheap synthetic materials in clothing, textiles and carpets. There was not the same effort, determination or finance available within the natural-fibre industry. The fact that this is changing is very heartening and rewarding for us as producers of merino wool.

In the last decade or so there has been a dedicated effort to promote and cultivate the awareness of New Zealand's quality merino wool through the New Zealand Merino Company, an integrated sales, marketing and innovation company which is

Me, Kate and Eric in the kitchen with Craig Adams of New Zealand Merino.

backed by merino wool fibre growers. What's given us hope, and faith, is being able to break away from the traditional and highly variable auction market for our annual wool clip. We now sign up for forward contracts and feel that we are part of a holistic system and industry based around the fibre, not just producers who supply at one end and then are ignored by the rest of the supply chain. Companies like Icebreaker have led the way. I remember being at the launch of the company and being so excited about the potential for wool in such innovative young hands, forging ahead with great new ideas for marketing and manufacturing.

Craig Adams, one of New Zealand Merino's marketing team, joined us around the kitchen table recently to update us on the company's marketing plans and new innovations. It's fairly obvious that New Zealand's high country adds more than a little magic to the marketing of merino fibre. But there's much more to it than that. New Zealand is recognised internationally as a very special environment for the growing of this fine, soft fibre. To capitalise on that, New Zealand Merino has developed product traceability, so you know which station your wool came from, and the 'Zque' mark, which emphasises sustainable management and good animal welfare.

Manufacturers — of wool carpets and wool textiles — ran their stocks down in the economic decline and currently a global shortage of wool also includes a worldwide scarcity of merino fibre. In 2011 New Zealand exports to China were record numbers compared with previous years. There is new optimism for the wool among merino farmers.

When our merinos are in the yards we get the opportunity to observe merino fleece up close. Carefully worked open, the

A 1911 wool sale receipt.

Account Sales 14 Bales of Wool Sold at Public Auction by Donald Reid & Co. Ltd. at The Otago Corn & Wool Exchange. Vogel St. Dunedin on 7th day of March 1911

3/1 439

by order and on account and risk of Mr Henry Scott

Glenorchy

Lot	Mark	Bales	Description	lbs Net	Price	£	S	D
58	HS BEESVALLEY	2	H Bred pt x B in No 106	519	10	21	12	6
59		1	Merino Pces	349	8¾	12	14	6
60		1	Merino & Bellies	362	7¾	11	13	10
61		1	Merino	348	9¾	14	2	9
62		1	X Bred & HB	229	8¾	8	7	
63		1	X Bred & 1st Pcs	281	8½	9	19	
64		1	1st HB Pieces	276	9	10	7	
65		1	2nd H Bred	260	10	10	16	8
66		1	Matted	272	9¼	10	9	8
67		1	2nd HB Pieces	282	8	9	8	
68		1	1st Mero Pieces	327	9	12	5	3
69		1	Locks	432	4¾	8	11	
70		1	HB Pieces & Bellies	419	5¼	9	3	4
			Black out No 10	45	6½	1	4	5
				4401		150	14	11

Charges

Railage	8	7	7			
Receiving, weighing & classifying 0/⅛d pr lb	2	5	10			
Repacking Bales at 5/ ea						
Commission 2½ per Cent	2	5	1			
Dissecting No 10 Bl		1				
				12	19	6
Net Proceeds £				137	15	5
To Credit of A/c				137	15	5.

E & O. E.

JH

Dunedin 21st March 1911

fleece reveals a thin dusty outer layer, below which is a dense forest of long crinkly fine fibres that are astonishingly white. The wool is protective, keeping the sheep both warm and cool depending on the season, and as an amazing bonus naturally produced oils also make the wool water-resistant. These are the attributes that make merino garments so wonderful to wear.

Merino fibre is fine and luxurious. Wearing merino garments is like being sheltered by your own micro-climate, keeping the body warm or cool and also offering rain shower protection. There is remarkable warmth even in fine, light merino garments, provided by air trapped between the fibres. Merino fibres have a complex cell structure that both holds and repels moisture, so garments offer warmth and constant temperatures even if the body has been sweaty or the garment has become wet.

Merino breeding has improved the stock. There was a time when the length of the fibre dictated breeding. Now it has been realised that the length of fibre is not the most important aspect of the breed, and fibres that are too long can cause processing problems. On Rees Valley we look for quality and quantity and the right length of fibre for processing. All wool is graded according to its micron; in short a measurement of the fibre diameter. On average, the Rees Valley clip comes in at 17–18.5 microns. Our wool is the sort of wool that will end up in luxurious, soft garments, worn next to the skin. Some of our stronger wool is sold to a company making fine merino socks.

One of Rees Valley's most exciting years for wool production was 1948, when the station held the New Zealand record price for wool sold in Dunedin. That year the station's merino fibre

---

Previous pages: Donald Reid & Co Limited wool buyers in Dunedin in 1948, assessing the Rees Valley clip that set a Dominion record.

set the standard for quality. A photograph of that day and the record wool clip sits in a frame in the hallway at the homestead, alongside the framed receipt for the sale and a strand of the illustrious 1948 fibre.

The farm diary, written by Jean Scott, recalls the excitement of the day:

> 19 April 1948
> *After dinner they went to the Head taking down potatoes, swedes and apples to send to Jean [Rooney, Doug's sister in Alexandra] by steamer. Great excitement when telegrams came from Beaumont saying 13 bales Rees Valley wool topped the market. A Dominion record.*
>
> *The news went over the wireless after 9pm. A great thrill.*
>
> 20 April 1948
> *Doug still taken up with the high price of wool. Hardly able to credit it.*
>
> *Several telegrams of congratulations and phone calls from locals.*
>
> *Information in the* Otago Daily Times.

Not only has wool undergone a renaissance, but old shearing methods have come back, too. Originally all sheep were shorn by blade shearing. Then machines were introduced. Now we have moved back to blade shearing again. We use shearing machines to take the wool off the sheep belly about three or four weeks before the main shearing. The machines are also used by Ronny and Tracey Hill's shearing team from Balclutha for autumn crutching and for eye-wigging most of the flock.

At main shearing the bulk of the fleece, which is obviously of higher value, is taken off by the same shearing gang using blade shears. It is an improvement in technique because the untidy bits off the bottom and belly have already been removed and so the process to recover the valuable wool is cleaner.

We are keen to make the most of our high-value wool and this technique assists that aspiration.

Blade shearing is highly skilled work and an age-old craft. Hand-worked blade shears allow shearers to leave a little wool on the sheep, as the blades do not trim so close to their skin. This is especially important for the September pre-lamb ewe shear, which happens when the weather is still cold. The remainder of the flock is shorn in October.

Eight shearers may work in the shed at one time with blade shears. Kate and I also use blade shears when we have a small hogget flock that needs crutching and for the autumn eye-clip. (Some merino grow excess wool and can get wool-blind. This wool is cleared away from around their eyes so they can see where they are going!) We work as a team and hand-clip and crutch each sheep one by one. The blades need to be cleaned and sharpened between each use, but the real art is placing each sheep in the holding strap so it feels secure and does not wriggle too much. Some do complain, of course.

And it *is* quite a process. Crunchy, frosty conditions mean waiting for the sun to come up mid-morning to ensure we are trimming and working with the sheep when they are dry. Often Kate will put an audio story on the iPod and we will work away to the dulcet tones of an author. On really cold days Kate puts the iPod inside a big furry, Russian-style hat to keep the battery at an operational temperature.

This year we listened to Christopher Paolini, and Bill Bryson expounding his tale of hiking the Appalachian Trail, *A Walk in the Woods*. The audio book lasted more than three mornings

Kate and me crutching and eye-wigging with blade shears.

of trimming and crutching. It wasn't too far removed from our conditions as he encountered snow and cold along the way. (Perhaps we are more fortunate here. Bryson walked the trail discovering the loss of certain species of tree and fauna due to acid rain. Some of these trees are present in New Zealand, including several we have planted at Rees Valley, such as hickory, hemlock, chestnut and elm. We've planted them for both shade and shelter, and in the case of the chestnuts, for food.)

Ling helps me move merinos in the yards in winter.

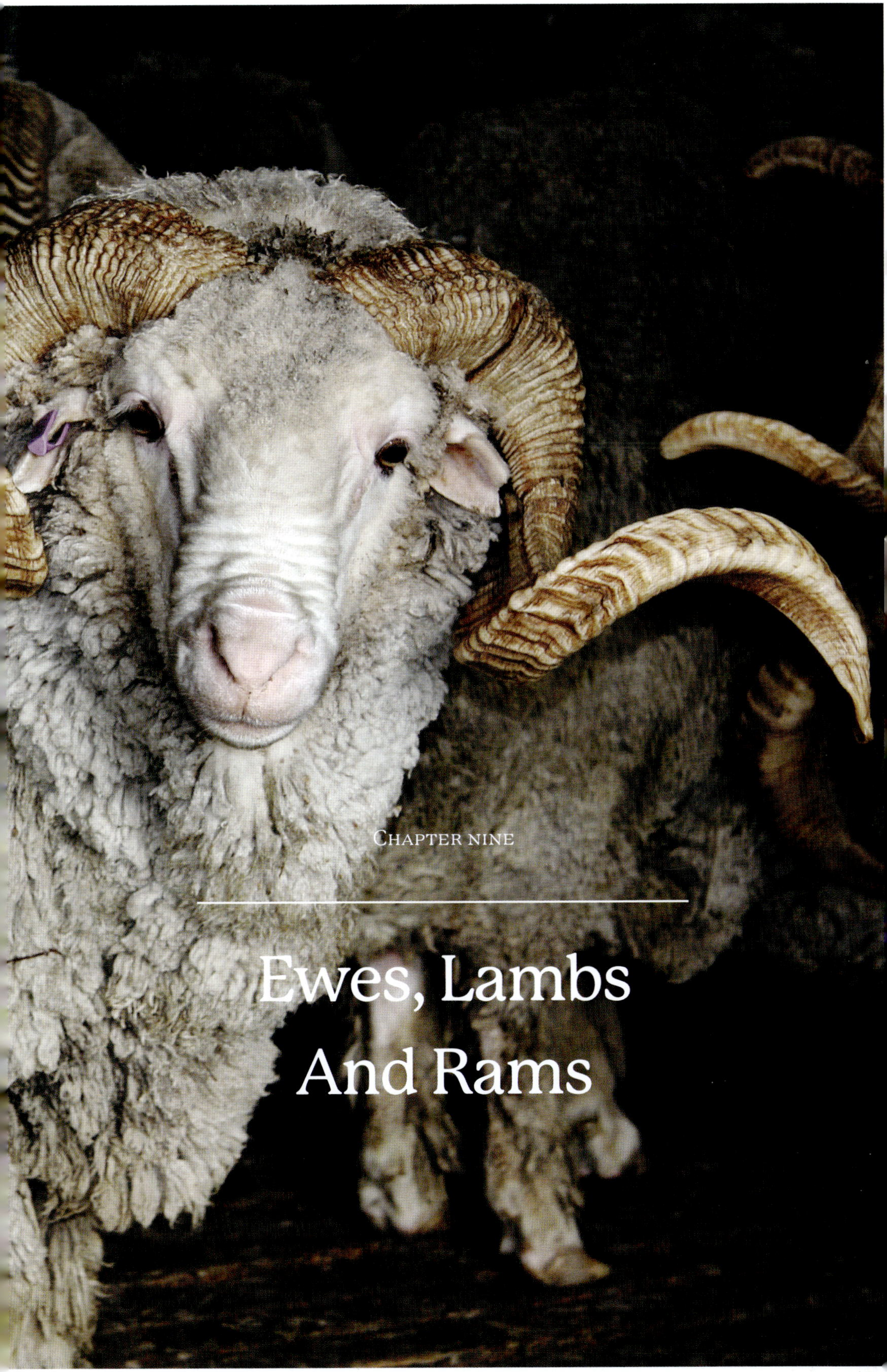

Chapter nine

Ewes, Lambs And Rams

I LOVE THE SMELL of the woolshed. Years of use have penetrated the floorboards and the timber pens, steeping them in the scent of wool which is so recognisable to New Zealanders, many of whom have spent time on sheep farms. This unforgettable smell is a treasured part of New Zealand's heritage. I'd been quietly optimistic when our most recent shearing had finished; the shed had been swept clean and I'd gone over to look at the big, neatly formed bales sitting in the woolshed waiting for the truck to come and pick them up. I hoped we'd get a good price but I tried not to think about it too much as I don't like to count unhatched chickens. So much was riding on it.

In late November, an email came from New Zealand Merino advising the price for our main-shear wool. Once our bales of wool had arrived at the wool store in Christchurch, a small sample of the fibre had been sent to Melbourne to the Fine Wool Sales and the results determined where our merino wool would be sold and used. It could end up in New Zealand in high-quality merino garments and textiles, or it could be exported around the world. The email contained encouraging news.

Ronny and Tracey's team had done a great job. Tracey is the classer and in charge of the shed; she's very skilled. And even though their team uses blade shears the rest of what goes on around shearing is a far cry from the original days of the run, when Mr Rees's sheep were first shorn outside. Nowadays our bales are carted away by truck and driven all the way to Christchurch on good sealed roads. In the early days of Rees Valley, bales of wool went by horse-drawn wagon, leaving the shed in time to catch the steamer down the lake to Kingston and then loaded onto the train and off to Dunedin. When the

---

Previous pages: One of our merino rams.
Opposite: Rees Valley Station wool from the 2011 main shear, waiting in the woolshed for baling and trucking to Christchurch.

road to Queenstown went in, it was still a rough ride from the shed out to the sealed road and on to Dunedin.

Just making the bales was an art in the early days. When I arrived at Rees Valley there was an old tramp press to consolidate the wool into bales. The press looked like two wooden boxes. Wool was tramped down in both halves, then these were put together and crushed down into a sacking bale. The bale was then hand-stitched along the top and marked. Each run had its own brand, hand-stencilled carefully onto each bale. Now, like most farms, we've got hydraulic wool presses. Simpler and easier, but still hard work. Now it's just a matter of a marker pen and a form to fill in, and each bale is digitally recorded.

While we admire the merino's ability to be independent and free on the mountains we are always mindful of the welfare of our stock. With the cyclic pattern of the seasons, and of the farming year, every summer the sheep will be out on the high, high country, with little human intervention. Over winter, we'll bring the pregnant ewes down closer to the homestead to keep their feed levels up and to support gestation. Later we will move them back up the valley for lambing. They don't need a daily lambing check like lowland sheep; merino lambs are small and the ewes lamb easily. I only need to check them once a week. In autumn, we bring them down off the high country, and they are carefully checked and crutched, and have their eye clip if necessary.

In winter, merino are put out on the winter spurs and the young ones also get a daily feed of hay, turnips or swedes. Obviously grass isn't going to grow much in winter, with its

---

The hoggets in a field of swedes, slowly discovering how tasty they can be.

morning dust of frost or snow, so consequently we produce additional feed to cover this shortfall. We grow oats to feed young sheep and the ewes around their shearing time, while we wait for the grass to grow on the lambing paddocks. Most of our hay is grown to feed the cows through calving.

Something that I learnt early on is that sheep need a bit of time to discover the sweetness of swedes. When we first put one-year-olds, the hoggets, into the swede paddock they walk all around the bulging plants and nibble only on the grass and weeds in between. That is day one or day two. Once the grass is gone they start to nibble on the green leafy bits, recognising they are edible. A day or two later the leaves are gone and a bare stalk sticks up from each swede globe. Then, out of desperation, the hoggets devour the stalks until these are gone. After a few days all that's left is the great globes of swede — something they haven't tried before. A nibble or two and then they realise they have left the best till last!

It's a partnership. We believe that if we look after our merinos well and provide for all their needs they will look after us by growing healthy wool. In contrast to fat lambs, which may be on a property for just a few months, each merino can become part of our station 'family' for what is a long time in sheep years. We get to know our flock closely and, yes, we can recognise and pick out individuals.

Every merino lamb more than a couple of months old has a Rees Valley Station ear-mark, the fork and a forebit cut in the skin of the ear. We count them, and can also draft them in the race: ewes are marked right ear, wethers are marked left ear. Originally this mark was all that was required; with modern technology we now use ear-tags as well. At one year of age each merino sheep gets its tag. There is a sequence of colours,

A hogget in the yards.

different for each year, so we can quickly tell which year each particular sheep was born, making drafting easy. Following the gender system, the tags are on the opposite ear to the mark: right ear for wethers and left ear for ewes. It's not quite like blue for boys and pink for girls, but it's a close second. A quick look at a flock and it is easy to see if there are any sheep out of place.

The calendar at Rees Valley always has a big ring around the date 15 May. For as long as records go back, that's the date, each year, that the rams are put out among the ewes. We work all the routines around this date. The timing is carefully managed for spring lambing (gestation is five and a bit months), to match the growth of grass for the lactating ewes and their lambs.

The rams are the aristocrats — the boys, the gentlemen, the studs — with their magnificent curled horns and handsome eyes fringed by a top lid of very fine, long eyelashes. In the mating competitions they usually clash and butt each other head on, but when Kate took the rams up the hill to the ewe block on a rainy day in May 2011, she came back saying the rams didn't seem that keen. Perhaps it was because they were toy-boys, many of them younger than the ewes. We had to rely on the girls to start the action!

We are moving the breed to polled merinos, which means that our rams, in future, will not have horns. We already have had some Nine Mile and Earnscleugh polled rams among a group of our ewes. The Nine Mile hoggets from last year look like a different breed, hornless and with more muscle. It will take a while for them to work through the flock and it is going to make dipping much easier as their horns won't get tangled in the apparatus! It's part of change afoot on Rees Valley. The next is to focus on producing merino for meat.

To date, merino meat has not been recognised as a specific and unique product by the market but this is about to change;

Silver Fern Farms, in conjunction with New Zealand Merino, is developing the market for free-range mountain merino meat.

We've eaten merino meat on the station for years. We cook different cuts, usually home-butchered, so we are able to experiment preparing and cooking across the full range of the animal. Kate really enjoys cooking merino meat and we often serve it for special occasions. She serves the leg roasted with all the trimmings. With any luck we have some left for sandwiches the next day. Hot or cold, it is delicious and easy to carve. Now top chefs and the food industry are using it, enjoying the tenderness and flavour. Merino meat has a different taste to lowland fat lamb. Merino are up and down the hills, and their meat is full of flavour, with a finer and denser texture than lamb. It's less fatty, too.

Producing merino meat is something we are looking forward to. Different skills and new schemes keep us on our toes. We try to be aware of fresh farming ideas and the farming papers keep us up to date.

Plum Sauce

This is a very old family recipe. Kate Scott used plums grown in the Rees Valley orchard to make this sauce.

6lbs plums
2lbs brown sugar
3 pints vinegar
6 teaspoons salt
2 teaspoons ground cloves
2 teaspoons ground ginger
2 teaspoons white pepper
2 teaspoons cayenne
1 teaspoon mace
2oz garlic

Boil all together till reduced to a pulp. Strain through a sieve. Bottle when cold.

Chestnuts and kiwifruit from the homestead orchard.

Chapter ten

Herding The Herefords

We have about 200 Hereford breeding cows on Rees Valley Station and the Fall Muster is a major part of our work with them. They summer with their calves up the valley floor to beyond the Twenty Five Mile Terrace from February to May, but once May comes it's time to bring them down to the homestead paddocks for the calves to be separated from their mothers and sold. The productive cows then winter on the blocks the ewes lamb in, tidying up roughage and pushing back the bracken. Importantly, we do not want stock out the back when the winter, and snow, arrives.

Kate likes to ride, so when she was working in Glenorchy we usually timed the cattle muster for when she was available and she could join the others on horseback. Ideally we need two or three musterers, preferably those with good horsemanship skills and also an understanding for moving stock carefully.

I find Herefords to be mostly obliging beasts. They know when the weather is changing, and as temperatures drop they will seek out the warmer valley floor or patches of hillside with warm sunny slopes. When we go out for the muster, many of the herd are ready to move and have even got themselves organised by the gate, ready to walk down the road.

I've done so many of these musters and yet I always look forward to bringing them out of the valley — not to mention the calf cheque that will follow. For the most recent muster Caro, who assists with our horses on the station, rode out, with her drover's roll behind her saddle, to the musterers' hut at Arthurs Creek the day before. She and her horse were thrilled to be part of the action and to have a drover's job to do. Caro is a great horsewoman, and very keen to be involved in anything on horseback.

At daylight next day, she rode around the perimeter of the

Previous pages: The Herefords cross a branch of the Rees River. The old Arthurs Creek woolshed is in the distance.

valley, gathering most of the herd together; however, we knew there were others still up in the hills and that we'd need to move on foot to bring them down. We went out in the Land Rover, hill-sticks in the back, so the musterers could start fresh for the climb. The sun shone and the tussock gleamed. Grabbing the hill-sticks, the team, with fit young legs, set off to look for the strays.

Some of the mustering team went up onto the terraces and knolls to check for stray cows and calves which had picked out some warm sunny slopes to 'hide'. The foot musterers knew the likely spots to look, and, indeed, a group of eight or more cows and calves were beneath one knoll hidden from view of the valley floor by the trees and a low spur.

Some of the cows have a deliberate ploy to avoid the muster while others are too busy eating in the sunshine to realise that the rest of the herd has moved. You can tell if they are being deliberately mischievous or whether they just haven't noticed the others have gone!

The task then is to get these last ones out of their tempting sunny pasture and moving along the route to join the rest of the herd. They can be unpredictable, and, having been on their own out here for a few months, they can be a little peeved that humans are trying to shift them someplace else.

It's a little more difficult to funnel a group through a narrow opening, and this day they chose the easier route and streamed off down the wide terrace, Eric on foot behind them, with just a few complaints and a few individuals deciding they might chance another side trip before joining the herd.

The cows swung along, keeping their calves reasonably close; all the herd was now on the move. The cows were by now five months pregnant with their next calves. Many of them had been part of the herd for years, as familiar to us as old friends. 'Come on, old girl. Get a move on,' I called as the Land Rover edged beside them. 'No need to go there. Where's your calf?' By evening they'd been on the move for a day and were pleased to get to a secure paddock with plenty of feed and shelter for the evening.

Day two, and the terrain was suitable for our horseback droving team to walk the cows and calves back along the valley road. The horses were caught and saddled up.

The drovers kept the cows and calves off the river and away from any chance to wander onto the adjacent station's land. Without fences, any sort of droving is challenging but the horseback team kept steadily at it, trying not to disturb the cows and calves too much. There were no dogs to help. We don't use dogs for cattle work; I've seen too many hurt by an unfortunate lash of a hoof and our cattle are happier without barking dogs nearby. And cows are honour-bound to attack dogs if they have calves at foot.

The herd was kept moving ahead and finally funnelled through the gate into another paddock for a second night. They were happy to settle down and feed. The horses and drovers were hot and tired but exuberant. Paddy, a horseman from Ireland, thought it was one of the best days of his life, in among this amazing scenery and riding for a purpose.

On day three the drovers saddled up for the final stretch along Rees Valley Road. There are a couple of bridges along the way and this always creates the most excitement. The cows would rather scramble down the bank and cross that way than be driven through the narrow, fence-like bridge area. They seem to have a fear of man-made structures, especially those that limit their options for getting away or fleeing from trouble. It's a natural instinct, I suppose.

Once they were securely in the cattle yards, we sorted calves and cows, then the heifers and male calves. The bellowing and mooing of ticked-off cows was quite something. But we had plenty of work to occupy us and their grumbles went unheeded.

All the calves had to have their ear tags inserted. The new requirement is for electronic tags that can be read by a scanner, and they take hours to fit. As I worked away I found myself muttering, Surely this was a bureaucratic decision made from a desk somewhere? Each one that is lost costs six dollars to replace.

After the muster, I was keen to get into the house to check the weather forecast for the next day. We usually like to hold the calves in the yards for a day before they travel by truck; the less food they have, the less waste there is on the roads of New Zealand. But I never like having to put the calves in the yards when it is very wet as it gets very muddy quickly.

Once the calves have gone the rest of the herd remains on the lower hill paddocks for the winter and during calving come onto the flat, where there are shelter belts to protect against the cold snaps that we always get in spring and where it's easier to feed out.

We have five Hereford bulls, and we usually buy a new one every year to introduce new genes to the herd. We keep the latest bull with a small group of cows for the first year to see what the progeny is like and also to check out how fertile he is. We like to see the characteristics of the offspring and what we are adding to the herd.

The glossy bull stud-farm catalogues are posted out to potential customers every May. Each bull has its vital statistics listed, in particular its estimated breeding value (EBV), one of the critical factors. If we go to the sales we arrive early to have a good look at the bulls. We will have made our short-list before the auction starts, instructing our agent, who

One of our bulls. I don't mind getting into a paddock with them. They are chosen for good temperament but we always treat them with respect.

will do the bidding for us. We look for Hereford bulls suited for the high country. Our criteria are good feet and a calm temperament — and then it's the largest bull we can afford!

The network of friends from Graeme's school days is still important to us and to our cattle operation. Waitaki Boys' High School has always had lots of farmers' sons on its roll and offered agriculture as a course. One of Graeme's great mates was Bill Simpson, with whom Graeme spent a lot of time outside the headmaster's office waiting to get caned; they'd keep a count with notches on their belts. Now Bill's son comes over from Oamaru once a year to look at our calves. Usually a stock truck follows to pick up those he has decided to buy. It suits both of us well.

And just as I am moving to polled merino, so I moved, not long after Graeme died, to hornless Herefords. Doug and Graeme loved the look of the horned beasts but I was the one who had to lean over them in the yard to test them for TB, and I had got pretty fed up with their horns and the bruises I got! Doug liked the fiery animals; I am all for cattle with a calm nature.

Following pages: Sorting the heifers in the cattle yards with Eric and Paddy.

Chapter Eleven

Tin Mills To Helicopters

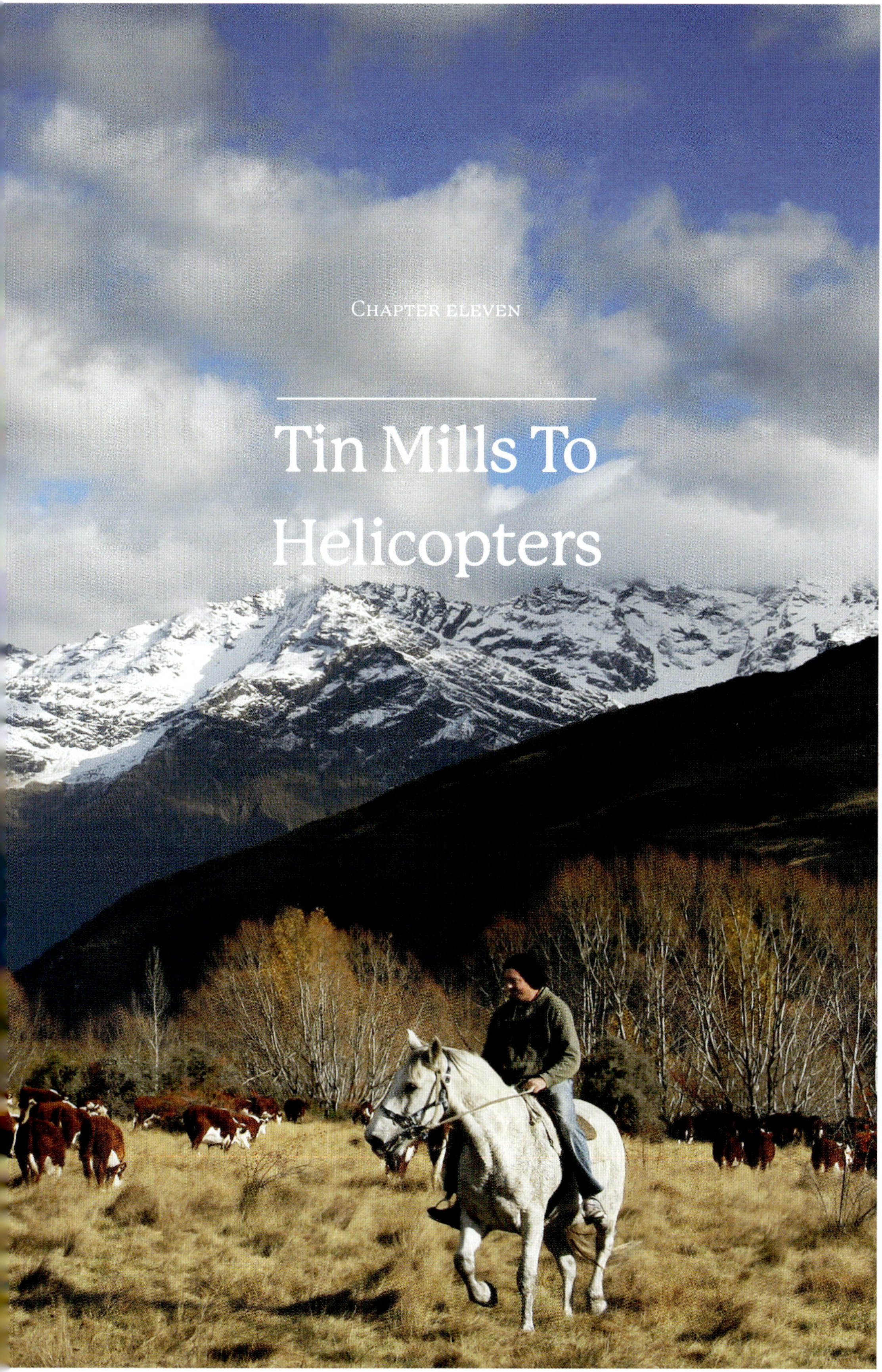

Fords and Rees Valley go way back. When it came to replacement time for vehicles, the Scotts were never in any doubt about what car or truck to buy. Ford. It was family loyalty. After all Jean, Doug's sister, was married to the Ford agent.

The Ford Custom that Jean drove to pick me up in Queenstown back in 1967 was followed by a Zephyr Zodiac, considered the upmarket version of the Zephyr and ahead of its time in shape and performance. Later Graeme and I had a sporty sort of Ford, which we named 'Bumblebee', as it was bright yellow with stripes. We sometimes had problems starting this vehicle — one time this happened on quite a critical occasion. I was about to produce a baby when the car wouldn't start. My brother — with me on board and the baby making moves to arrive — towed the car around the yard until it fired and we could head off to the maternity hospital in Queenstown for the birth.

The Rees Valley truck — a veteran, venerable, hard-working 1950s Commer Super Poise tip-truck, to be precise — is called Henry, a hat-tip tribute to Ford history, even though it's not actually a Ford!

Mechanisation changed farming as tractors replaced draught horses in the 1940s. The first tractor arrived at Rees Valley in 1947. Like our series of Henry trucks, every tractor was given a name, the first two being named after the radio comedy serial *Gert and Daisy*. What was the connection between a radio serial about a Jewish family in New York and a farming family in Glenorchy? The Scotts listened to it in the evenings, and the name struck a chord.

The arrival of Gert was a big day on Rees Valley. The farm diary records:

Previous pages: Pride and Paddy move the Herefords down the valley. Opposite: The Commer truck called Henry, in service since the 1950s and still going strong.

REES VALLEY
STATION
GLENORCHY
TARE
14.2

Little Liz

...t Gorge Gate

29 November 1948
Arrived with our latest purchase Fordson Major tractor on ½ tracks

All hands were busy converting from rubber to steel

30 November 1948
Held public demonstration about 15 attended

The arrival of the first tractor provided the opportunity to re-establish the track up the valley, once used by goldminers and hardy tourists travelling by horse and trap. Courtesy of a blade attached to the front, the tractor restored the track, later making it possible to drive in by truck, although it was still necessary to get through many fords. Needless to say, the first 4WD Land Rover to arrive at Rees Valley Sation made this journey so much easier again.

The names of the tractors that followed Gert came from popular cartoon characters. Our two green Deutz Fahr tractors, still very much in use today, are named *Kermit* and *Robin* (Kermit's nephew) from the TV series *The Muppets*. Kermit does great service despite being named after a frog. Robin, the smaller tractor, tows the hay feeder during winter on a daily basis. Our hay feeder is thirty years old but we still like it because it allows us to take a slice off the hay bale, a wide slice for bulls and a narrow slice for sheep.

After the war, many vehicles found their way to farms

Previous pages: A snapshot of Little Liz, loaded up with dogs, from the Scott family photo album. Following pages: Gert the tractor and 'the big truck' parked outside the Arthurs Creek huts in the late 1940s. Once trucks arrived it was no longer only a horse that could get you up the valley.

throughout New Zealand. On Rees Valley, Little Liz the pick-up truck was a step up from horses and the pack horse and the forerunner to the Land Rover. In the early 1950s, it was an hour to Arthurs Creek by Little Liz. The road has always been unsealed and with fords to cross; but even with these challenges Little Liz was so much faster and easier than horseback, which took a good two hours or more. On the back of the truck, the dogs could arrive fresh and ready to go. The advent of Little Liz and the trucks that followed her meant that family and friends could go to the back of the station more easily, too.

We've got the odd relic or two sitting around, including an ancient tin mill. Looking at these old pieces of historical equipment it's not hard to see how they work. The first steps in farm mechanisation were simple machines with cogs and wheels. The tin mill was used for threshing oats, and the straw it produced was then put through the chaff cutter. It is the oldest piece of machinery still on the station, although it's no longer used. These days it sits behind the equipment shed, a relic of an earlier time that will, we hope, find a new career as a museum piece.

All three children had ponies but very early on Eric preferred tractors. He would sit on Graeme's lap and learn how to drive the bulldozer and the digger. His legs were too short for any equipment that had pedals but those with hand controls were easy for him.

When Eric was twenty, he studied for his helicopter pilot's licence under Simon Spencer Bower, considered one of the best trainers in New Zealand at the time and recognised as the world's top R22 pilot, with over 10,000 hours on a Robinson R22. Eric's training was different from many others as he had to fit it in around farming. His practical was at Wanaka

Gerr & the b
at Athers

g Truck
Creek.

Helicopters, and theory was studied through the Open Polytechnic correspondence courses.

The tiny yet practical R22 Beta II helicopter we've had since 2009 performs an outstanding number of jobs around the property easily and quickly. One of the big helicopter tasks is mustering the sheep off the high mountain blocks. At first, we'd fly the musterers out with their dogs to the tops. It would be a big day, with several helicopter trips needed to get everyone on the spurs. Different teams would be left on different spurs, from where they would work their way down, gathering the mob as they went. It wasn't always straightforward; one of our home movies shows the dogs and musterers waiting for fog to clear so the helicopter could get in to pick them up.

Now we've moved to fully mustering by helicopter. It's a task at which Eric has proved expert, and our sheep have now got used to the idea that the helicopter buzzing close by is there to move them down. Any helicopter pilot who undertakes mustering has to be familiar with and accustomed to stock work. Eric is experienced with stock, understanding when to put pressure on animals as they move and when to hang back. It's critical. It's also essential to know the lay of the land and the tracks where the animals can safely move down to the valley. If you put pressure on at the wrong moment the sheep can go the wrong way, stream back up the hills or get into difficulties.

When done with skill, heli-mustering is a much easier process than the hard yards of doing it traditionally. That involved waking before dawn in one of the huts to get up to the tops first thing in the morning, hours of foot slog uphill for the mustering team — men and dogs — before their day really started. Mustering with just one man in a helicopter is a logical step although, as Eric says, something is missing compared with the traditional muster.

'As a boy I remember the crowd around the dining table after the end of the big muster,' he says. 'The days when the men and women and their dogs from the local farms were all gathered together in the valley before setting out to muster their allotted block for the day. I remember it being such a big deal, the muster. Everyone would go out and do the job and then come back and there'd be a big dinner at night. There were a few beers, of course. It's a wee bit of a shame that this walking and work with the dogs is now disappearing from the annual life of the high country.'

One of the benefits of heli-mustering is that fewer sheep get left behind and we do not have the straggle muster, looking for those still on the hills, as they used to in the old days. Eric can do the whole job moving mobs of merino down into lower blocks. He can also put the helicopter down, get out and open or shut the gate, or whatever. You can waste a lot of time hovering around just making sure the sheep keep moving so he often gets them moving, mobbed up and ready to go, then we put people out with the different mobs to keep them heading down. This cuts down the flying time.

The helicopter is also useful for easily moving things around on the property. We can fly people or feed out to stock. In 2011, when the heavy snow came, for example, we had already moved our stock down while other stations spent days with hired helicopters getting their stock sorted and out of the snow. Salt, put out on the blocks to replace minerals that are not found in the poor soil, can now easily be taken out by helicopter. Eric also uses the helicopter to shoot feral goats as part of our pest control programme. He flies and a mate shoots, and they always take care to ensure that the goats are dead before they move on.

Of course, there are risks in using helicopters on a station. The weather in the mountains is so changeable and can be so severe. Weather checks before flying are made hourly, and more

frequently prior to setting off. Eric usually looks ahead to the weather patterns for the most suitable day for flying.

A new innovation is Spidertracks, a live flight-tracking program that enables the helicopter to be tracked on a website in real time. The pilot switches on the tracking device as he sets off and the homestead can log in and actually see where the pilot is and where he is working. As well as the ELT beacon in the event of a crash, it is a security measure. Spidertracks also has a pre-set button for messages which can be used in an emergency. Just press one button to alert others to a problem. I find that very reassuring.

In 1988, Doug, my father-in-law, was filmed for the TV show *The South Tonight*. As he answered Jim Mora's questions, it was clear the mountains were his natural habitat. 'You can't retire from the mountains,' he told Mora. He also observed that with the use of helicopters something had been gained and something lost. 'It puts you on the job where the sheep are rather than the time and effort of getting there. For the straggle muster you can see where they are,' he noted.

But, he went on to say, it 'took away all the camaraderie of the mustering and camping together'.

In the early days, Rees Valley employed mustering teams who would work around different stations, a bit like shearing gangs. There was a tale from these times, about a bit of heavy drinking and how some of the musterers got the worse for wear one evening. When the Rees Valley head shepherd found out, he ordered the group to be at Stony Hut, at the head of the creek that drains into the Shotover River, by morning — a walk of five to six hours (in daylight) up steep hills and including a

---

Previous pages: Eric takes off.

climb over Dismal Saddle. The effects of alcohol and the dark would have made this quite a journey!

The methods of farming may have changed but then as now, drinking on the job is a no-no!

Chapter twelve

Snow Business

INVINCIBLE — THE NAME HARKS back to the goldmining days of Rees Valley. On the station we have Invincible Creek, Invincible (or Reef) Spur and the Invincible Mine, so creating the Invincible Snowfield on Invincible Spur seemed logical. Also, it's rather a cool name for a ski field. As it turned out, coming up with a name was the easy part!

Some years back there was a strong push for farmers in the high country to diversify into tourism. During that push we were approached by developers who wanted to create a ski field on our property, which would have involved building a multi-million-dollar road up to our good bit of snow and turning it into another commercial field.

Our reaction was, absolutely not. Our approach is to lightly touch the land and not to bring in the heavy artillery of bulldozers and roading that would change it forever from its natural state. However, we were aware that up in the Richardson Range we had a particularly nice bit of snow that had been used for heli-skiing and heli-snowboarding from time to time. We liked the idea of the traditional ski-club fields which used to dot the New Zealand mountains. They have a low-key atmosphere: just a simple hut and a rope-tow and groups of friends enjoying the snow together. It is a relatively benign use of a resource. And so the idea of the Invincible Snowfield, to be accessed by helicopter for adventurous skiers and snowboarders, with skill levels of intermediate or above, was born.

At no stage did we think we were going to make pots of gold; in fact, the returns have actually been a bit like goldmining — an awful lot of effort for an occasional, or non-existent, glimmer! Initially the plan was for a partnership.

Previous pages: Wind whips over the tops of the Forbes Range from the West Coast. We always know what the weather is like on the other side of the mountains.

Above: Invincible Hut on the Invincible snowfield. Below: Skiers and the rope tow on the snowfield on a perfect day.

We'd provide the site and once we'd installed the 700-metre nutcracker rope-tow together, the partners would be responsible for the snow activity. Eric took a Mountain Safety Council avalanche course and we built a hut, a snug wee building that sleeps ten people and is heated by a woodburner. Cooking is done on a gas cooker. Set at 1650 metres, the hut has fantastic views and a little veranda to sit on to enjoy the scene.

And how wonderful it is to be a heli-skier on a sharp clear day just made for skiing, being whisked by chopper up to the field, ready for a day's action on a snowfield that is exclusively one's own. The rope tow gives 300 metres of vertical lift per run — once you master the challenge of rope tows! It's powered by a Ford 2000 tractor and takes skiers and snowboarders up to 1800 metres to begin their runs.

The hut is used in summer too. Groups can be zoomed up there by helicopter, avoiding, if they want to, the four-and-a-half-hour walk from the valley floor. They can be taken straight into an alpine environment, to either walk up to the peak and out to Reef Saddle, from where there are spectacular views into the Shotover, or enjoy the alpine flora, especially in summer when so many mountain plants are in flower. We don't hire gear out but we can arrange catering and we fly in all the firewood that's needed to keep that woodburner going and the hut warm and cosy.

Several crews have filmed up at the ski field. On commercial fields, such as Coronet Peak or The Remarkables, there are punters all over the place. But up at the Invincible Snowfield the crews know there is no one who could possibly ruin their filming. Indeed, location scouts are regularly checking out Rees Valley. They usually have specific ideas of where and what landscapes they want. If our property suits we tell them to go for it. Our only stipulation is that the product being advertised is 'respectable' — no cigarettes or irresponsible use of 4WDs allowed!

Much of what's filmed on the station is commercials or stills for overseas clothing catalogues. They usually want access to landscapes with waterfalls or bush or glaciers in the background and tussock in the foreground. We can easily oblige! Filming is unpredictable; some years no crews come to the station and other years we might get three or four projects. We try to keep out of the way and not to drive mobs of sheep through the middle of a set!

Tourism New Zealand regularly features this region in their publicity. It is recognised as some of the most magnificent mountain and lake scenery in the country.

The ski field development in the end was a bit of a salutary lesson for us. The partnership didn't work out and the costs — and the pitfalls — were so much higher than we had anticipated. We do need to diversify but to do so you need lots of time, and very deep pockets. It was a big distraction from the business of farming. Meantime, that powder is there every winter.

Mulled Elderberry Wine

We pick elderberrries from the trees that grow wild on the station and around the district to make this wine.

1.25kg elderberries, weighed without the stalks
4.8 litres water
1.5kg sugar
5 teaspoons yeast

Crush the berries and pour boiling water over them. Stir well. Leave to soak for 12 hours. Crush well once more then strain through fine muslin and put the juice through a jelly bag.

Bring the juice to the boil then turn off the heat at once. Pour into the fermenting vessel and add the sugar immediately, stirring till it is all dissolved. Allow the brew to cool and then sprinkle the yeast on top and stir in.

Cover and leave to ferment for a minimum of 14 days, after which proceed to bottling. Fermenting can take much longer. In any event, do not bottle till the wine has worked out completely and there are no more bubbles rising, otherwise your bottle could explode.

To mull: pour 2 bottles of wine (about 1.5 litres) into a large saucepan. Add 1 cup of sugar, a few whole cloves, 1 cinnamon stick and 1 sliced orange then warm through to dissolve the sugar and infuse the spices.

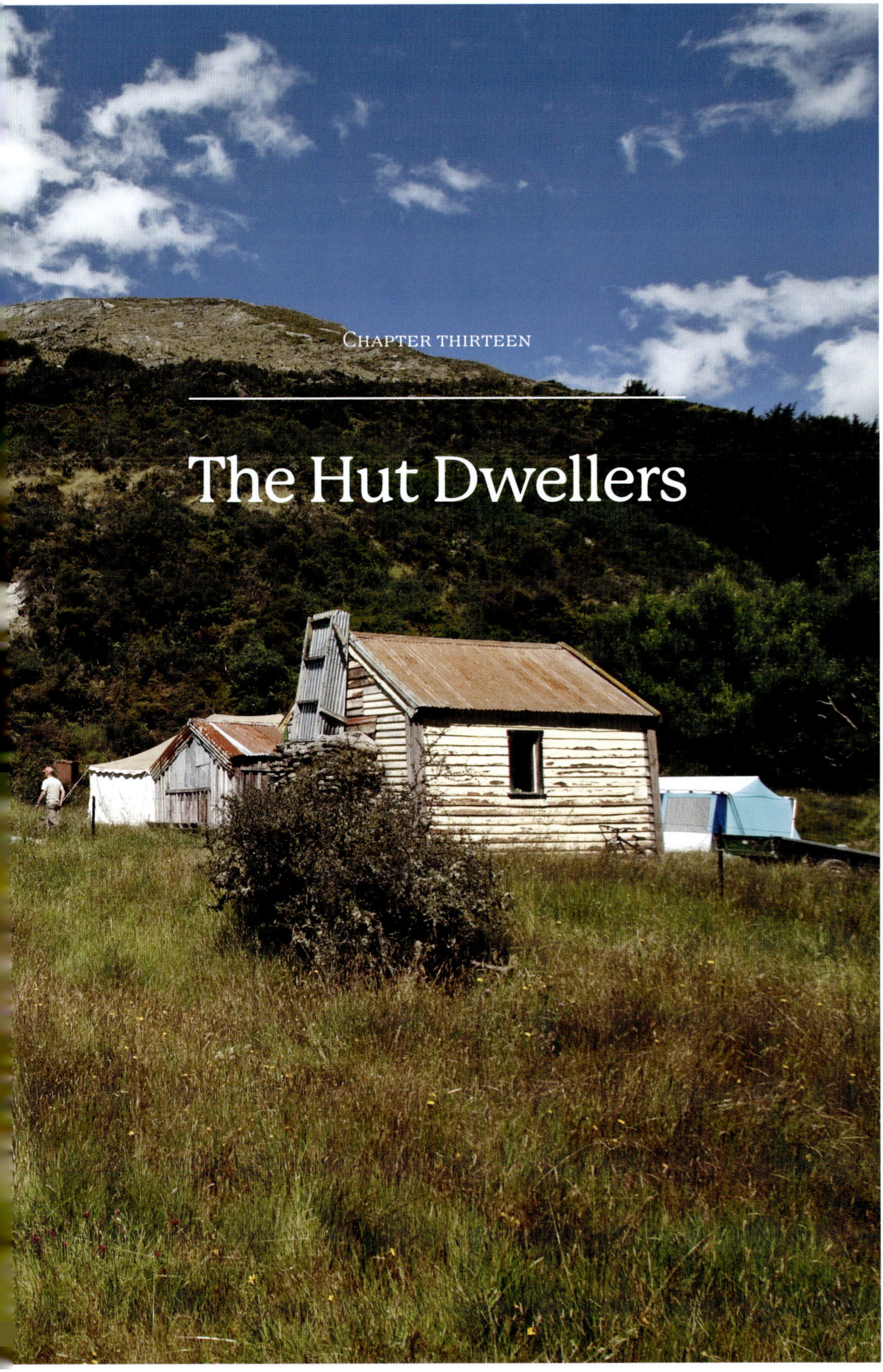

Chapter thirteen

The Hut Dwellers

Beech sapling frames. Corrugated-iron roof and walls. Dirt floors. Tea chests and apple boxes for furniture. Jars and bottles as candle holders. No electricity, no phone, no TV, no radio, no road access. The dust of time and the patina of use by musterers, boundary keepers and possum hunters for over a century . . . the moment I open the door of just about all of our mountain huts, I am reminded that they are special buildings.

What is it about that simplest and most rustic of shelters, the back-country hut, that makes the experience of spending time in one such an indelible part of being a New Zealander? Perhaps I am mythologising. That's changing, of course. School groups often used to have a week in the bush or out in the remote country. But now, it seems to me a great shame that the huts' honesty and genuine naturalness, as well as the ethics of their use, and the knowledge of back-country survival, are becoming lost to newer generations.

Our family has a special hut, built by Graeme before I knew him. It's our crib, our bach. It's our place remote among the mountains, without electricity or any modern facilities. We love it. It has been the place of so many our family holidays and it holds our family memories. The fact that Graeme's ashes are now there make it even more dear to us.

Our going out to the hut was always linked in with the stock work. In summer, after shearing, we'd take the sheep part of the way to the summer grazing the first day and then stay in the hut. The next day we would take them a bit further into the hills and then go back to the hut to stay for another night or more.

We'd often go to the hut on New Year's Eve, not to party but instead to get away from the need to party. We're not party animals, so spending New Year's Eve in one of the back-country

---

Previous pages: An Alpine Club members' camp around the old Arthurs Creek huts.

The children and I walking back down from our family hut on Big Devil Spur in the 1980s.

huts was a chance to enjoy peace and quiet and start the year in our own way. The children would play on the rocks and create their own imaginary adventures. They were also learning about the hills and the mountains and how to use their own hill-sticks, the straight pieces of manuka stripped of bark, that are essential high-country equipment. Hill-sticks assist walking up and are especially indispensable for getting safely and securely down steep hills.

The most distant hut in the station's back country was the Head of the Rees Hut, later dismantled when Mount Aspiring National Park was established in 1964 and it was found to be on the wrong side of the park boundary. We have early black-and-white photographs of that hut, which show the men and their working dogs and horses outside.

That old hut was an essential shelter for musterers working the back country, in particular on the first night of the Fall Muster, the big autumn muster. Next morning, they'd be on the move, mustering on the furthest spur and then heading to the next hut, and so on along the run. The huts were built in sheltered areas, near water and bush (for firewood). They usually had dry kindling and firewood, bunks and basic cooking and eating equipment. The shelter they gave was critical at times; there is no doubt that these huts would have saved lives. The huts were also a place of camaraderie, where the men yarned after a day's work. Photographs show their sun-worn faces and their clothes tell of hard work outdoors.

Another hut, for trampers walking the Rees-Dart or heading over the Cascade Saddle to Wanaka, has been built in the national park in a similar location to the original Head of the Rees Hut. Its name, Shelter Rock Hut, recalls the very earliest days of the run, when musterers headed for Shelter Rock, a large overhanging rock face that offered the most basic of cover for the night and protection from the weather.

There is an unwritten code of ethics for staying and using back-country huts which is now being ignored. I find it is so disappointing that reckless and irresponsible people abuse remote huts on our land. Locks have been broken and gear removed, without any care for the fact that others may need to take refuge there. One of our huts was once completely cleaned out. It is a huge disappointment to realise that modern itinerants, who can now easily get into remote areas, have no respect for or understanding of the basic survival essentials we leave for those who might need them.

If people break in how do we know they are there? Sometimes poison has been laid for pests or hunters may be working there, endangering them. If they are hunting themselves they may endanger us or the stock. The unwritten code has always dictated that people using back-country huts leave fresh, dry kindling and firewood, and also leave the axe provided in case any more firewood is needed. People who know the back country well understand the protocol. They ask us if they can use one of the huts. They *ask*. A simple request but important. I would hope that they would have mountain rescue radios and I always watch the weather. If I know it's going to make a turn for the worse I'll let them know. You don't mess with this country, you heed the rules — written and unwritten. Rising rivers are more likely to kill than big snows here!

Recycling is part of the life at the station and the family and I are really dedicated to doing our bit for the planet. However, it isn't just the daily waste and rubbish that is recycled; the process and philosophy is much more deep-seated. The chooks get kitchen scraps. The dogs get other bits. That's pretty straightforward. What's less obvious to outsiders is the total reuse policy that's traditional right through the high country.

s Wilson. Cr
ead of the Re

ossing the 25m

s Hut.

Back-country huts and shelters have long been rebuilt from materials that were available close by; anything and everything was reused, and then reused again. Any other materials had to be carried in on pack horse. It was no easy task to get these additional materials to the site in often mountainous and remote locations, so back-country hut construction was based around what you could scavenge nearby. The huts on Rees Valley were built reusing materials left over from the mining era. The Invincible Mine post office, for example, had a schist and tin chimney. That makes it sound grand; it was, in fact, really basic — one room and a fireplace. Long after mining ceased, some walls remained upright as did the metal fireplaces. These later became the base of a trampers' shelter built inside the original stone walls and using the same fireplace and probably reusing the roofing iron.

Two small huts were combined decades ago to make the shearers' quarters and only now are the days of this particular building numbered. The remnants of those old huts provide memories of those who have travelled or worked this land before. The livery stables James Dunery built, which later became the first woolshed on the station, recycled many of the timbers, cut from beech trees on the property, from the time of the mines. The building remained standing long after the new woolshed was built; however, now it has sagged and fallen over.

On other buildings, a door or window may have come from another structure, bits of miners' shanties have been reused, and anything that looked useful repurposed. Past generations didn't buy new; it was then far too difficult to get it here.

Previous pages: Another great shot from the family photo album. The Head of the Rees musterers' hut was demolished in the 1970s because it was inside the boundary of the new Mount Aspiring National Park. Opposite: The old yards and woolshed at Arthurs Creek.

The sheds at the hub of the station house tractors and trucks and tools, and every bolt that comes off them and every piece of wire is kept if possible, in case we find another use for it. As I tell Kate and Eric, after all, the hardware store is a couple of hours' drive away. They are, however, I have to admit, a little sceptical about some of this! Visitors may look around our yards and see the sheds containing old machinery and parts, the long-unused trailer, pre-used fencing wire and lengths of timber. It looks a little like a hoarders' yard to the inexperienced eye. However, other farmers will recognise the wisdom at work here. We know what things we have and, when needed, they are there ready to use.

When we build or plant anything we expect it to survive for a long time. We expect a fence to stand for fifty years at least, and some of the fences on Rees Valley Station are over a hundred years old. We like to think that when we replace these, the new fences will last just as long. The early fences were often built not to manage stock but to protect gardens and crops. There was a sawmill at Kinloch and timber was easy to obtain nearby or on the property itself. Huge trees were milled; on Rees Valley mountain and red beech was used as pit props in the Invincible Mine.

Around the woolshed, there is red-beech post-and-rail yard fencing that is one hundred years old, cut from the trees on the property and hand-milled without machines at the Kinloch sawmill. We admire the workmanship which carefully shaped each rail, cut on an angle, in cross-section like a long triangle — thin at the top, then wider and thicker at the base to allow the rain to run off. These old yards are still in good working order and are part of our philosophy to reuse and reuse again. The final duty of the fenceposts may be as firewood.

The old shearers' quarters.

Chapter Fourteen

Horsewoman

MYRRA WAS MY DOWRY! I brought my own horse, my riding skills and cash from the proceeds of selling my car to my newly married life at Rees Valley Station in 1971.

How did I start riding? When I was growing up anybody who went past on a pony was like a magnet to me. My parents encouraged my interest, and one day we got talking to Beverley Orr, a girl who had a couple of ponies that she took to shows. I would ride one while she rode the other, and then I would help her with the grooming when she went to the shows. I guess she really taught me to ride.

It must have become obvious to my parents that I was not going to get over this pony phase very quickly. So they bought me one and I started going to pony club. We rode from the back of Manurewa, up Hill Road, through the country and down Redoubt Road. It was quiet countryside then but these days the area is built-up and part of the suburbs. I continued riding right through my university years, and when I got my first vet job in Tauranga, Myrra came too.

Myrra was an Anglo-Arabian. She was originally called Grey Mist but there were about ten Grey Mists at the pony clubs I belonged to, so I changed her name. In any event, she wasn't a misty sort of personality but instead rather dashing and lively.

As our wedding day neared, Graeme borrowed a horse float from Temple Peak Station and trailed it all the way north. After our honeymoon in Northland we went back to Drury, picked up Myrra and then trailed the float with her on board all the way south to Glenorchy.

For the first year or so I was able to ride often around Glenorchy and in the Rees Valley. Being able to ride freely

---

Previous pages: The lower reaches of the Rees River, not far from the homestead and looking south towards Glenorchy, with the Greenstone Valley about the middle of the Humboldts in the background.

through the valley was wonderful; it was invigorating to be out in wide-open spaces, riding through the beech forests and into the tussock country out the back. These days there are horse-trekking companies in this area and it is easy to see how much their clients enjoy riding in these wonderful surroundings. One trekking company occasionally takes groups up through the station to stay in one of the back-country huts, an unforgettable experience.

There have been horses in this area from the earliest days of Pakeha settlement. The first European settlers arrived on horseback, so working horses came first, as pack horses, then droving and draught horses for ploughing and cutting hay. Teams of horses were the tractors of the day, driven by 'teamsters' skilled in getting them to work hitched together, pulling heavy equipment. They were followed by horses ridden for pleasure.

The Rees Valley photograph albums brim with images from the days when horses were the sole means of transport and absolutely critical to farm work: pack horses loaded with swags and gear, drovers saddled up. It seemed a whole army of people were employed around the horses, including the teamsters and blacksmiths. These photographs spell out the history of farming life in the high country. The arrival of tractors and other vehicles changed all that.

Horse-mounted troops were a key part of the early days of the First World War and many horses were sent by ship to accompany their riders. Farm horses, well used to human handling and the rigours of work, were considered valuable and useful, ideal for the war effort. Henry Scott gave three of Rees Valley's farm hacks to the war effort, as an old postcard from 1915 that we have kept reminds us:

From Rees Valley Station
February 22, 1915
To Master Douglas Scott, Stafford Hospital Dunedin

Dear Douglas
I think these three horses must be Barney, Bruce and Monk that Daddy sold for the War.

Love from Mother and Jean

We do not know why Doug was in hospital at such a young age. Perhaps to have his tonsils out? In 1915, he was five and would have been sad but proud to see some of the farm horses go to war.

Riding parties have historical significance and were an early part of tourism in the Glenorchy area. Paradise House and other hotels used to take people on trips, either on a horse, or in a gig or spring cart. The wonderful photographs of these days capture the romance of slow travel, when people turned out in their finery even if journeying by horseback!

Riding parties are still part of recreation in Glenorchy. Many of the local people have leisure horses which they enter in shows and eventing competitions, here and away. There is a strong pony club and a riding club. The riding club is casual but often enthusiastic members may set up a riding trip and all, including young people, are welcome to join. I try to go when I can.

However, now I hardly ever have time to ride for pleasure. When there were two working men and a couple of young apprentices on the station assisting with the work, I had more time to go for a ride on my horse. But when we stopped employing people I had more commitments. It's not from

---

Saddling Pride for the cattle muster.

Me on Jerry in about 1988. Note the neatly trimmed holly hedge in the background.

choice that I ride a quad bike but out of the need to go further, do more and faster, in a day.

In my ideal world I would have somebody working with the horses. They would arrive at the back gate at nine o'clock in the morning with my horse already groomed and saddled. I could just get on and ride round and look at the stock and everything else to my satisfaction, and then I'd just hand the horse over to the groom. It's wonderful to dream. However, horses are less expensive than quad bikes. With the price of fuel and bike expenses, horses as transport may come back!

My current horse is Noddy, a retired trotting horse. And trotting up to the fence whenever I go along the farm lane is a family of little black ponies, which we didn't expect to ever have. It all started with a search for a pony for Kate when she was young. We looked everywhere for a suitable pony that could be ridden bareback and which could provide a real riding experience, a pony that was for the pleasure and the fun of riding free in the wide-open country.

We didn't think we would need to break in a pony for Kate but that is what we ended up doing. Morning Star was, unsurprisingly, born in the morning and had a star on her forehead. We actually nicknamed her Donk, short for donkey, because she was such a quiet thing as a young horse. She was easy to break in, friendly and sensible, but she was also a half-thoroughbred so she could actually move quite fast and became a great little stock horse. Kate went horse-trekking and also did stock work with her, as Glenorchy was not serious pony-club territory back then.

Diane also had a little Shetland pony called Johnny. Johnny was a little terror, who would bite, hence his nickname, 'Sandfly'.

Because Morning Star seemed to be turning out so well we thought it would be nice to breed a few more ponies so other children could have the same experience. I found a stallion,

Matai, locally, who was only about 15 hands high and had a particularly nice temperament. Ideal. We brought Matai to Rees Valley and started trying to breed little ponies. We nearly gave up because he was here for quite a few years before anything happened. Then, finally, Eric's pony Denny produced a foal somewhat unexpectedly.

Now we have a whole family of little black ponies. They are an ideal kids' pony, pretty but also hardy and with good natures. They've all got a sense of humour and a great sense of mischief, which can be a challenge, but that's just the sort of pony that kids want or need to learn to ride on. Children can begin to understand the personality of the horse and have a lot of fun with them without having to be competitive. Our black ponies are ideal for pony camps and a couple of them have been away for a year or two to a family. I always tell these folk that when their child has grown out of riding or is no longer interested, the pony can come back home and rejoin the others in the family.

Most of the kids' ponies that I've seen in the last thirty years have been bred up and trained with a view to being sold on as a valuable show pony or competitive horse. I don't believe that's what kids' ponies are all about. I had so much fun riding my pony bareback when I was young, thundering through the gum trees on a reserve or taking him for a swim in the sea. Riding for pure enjoyment is an aspect of rural life and pioneering New Zealand history. It is our heritage. It reminds us of back when children rode their ponies to school. This experience is worth preserving.

Morning Star only died last winter, having been retired for

My current horse, Noddy, with his shaggy winter coat. Following pages: Miro and Ceilidh. Miro is one of our long line of little black ponies, which all have Maori, mostly botanical, names. Ceilidh is one of our half-thoroughbred ponies, who have all had Gaelic names.

quite a few years. Kate continues to ride when she has time. She enjoys assisting the cow and cattle musters on horseback so we plan the timing of that around her other working commitments.

Like many other enthusiastic riders, I've taken part in a few days of the Otago Goldfields Cavalcade, a trek throughout the region, following the old gold trails. I'm always at that notorious horse-riding event, the famous annual Glenorchy Races, the country horse-race day that's Glenorchy's biggest local celebration and charity event and which draws crowds from all around the region. It's held on the first Saturday after New Year, and was started in 1962 by the local football club as a fundraiser for the community. It is held on what is now also the golf course! Where once I was a keen competitor — I once won the Stockman's Race on Myrra — nowadays I attend as the official vet.

There are several novelty races such as the Saddling Race, in which riders ride bareback to the end of the straight, saddle their horse with the saddle that's waiting there, and then race back again; Double Banking, in which the rider races to pick up a partner at the end of the track, pulls them up to sit behind them, and then goes hell-for-leather back down the track again; and the Walk, Trot and Gallop, in which riders walk a horse partly round the track, then trot for a section and then finish at full gallop. The programme also includes the Trotting Cup, the Galloping Cup, (the marquee event of the day), the Ladies' Gallop and the Locals' Gallop. There is also a relay race with teams of four riders, some from different stations, others made up of horse trek guides and anyone else raising a team. It's become a reunion day for Glenorchy people because everybody who's ever lived in the village or the wider district knows when race day will be and they always turn up.

Outside the homestead in the late 1970s, with Kate and Diane on Shiloh.

The organisers were a bit of a chauvinistic lot in the early days. Evidently they didn't want 'their' ladies to be injured, so they had a rule that the only race that women could ride in was the Stockman's Race. Why? The Stockman's Race involved a race around hay bales. These didn't hurt if you knocked them over. The problem was that most of the horses around the district were owned and ridden by girls. They'd have to find a bloke to ride their horse on race day in other races than the Stockman's Race. Problem was that race day was a day of some celebration. The riders had often consumed a few beers and were therefore often not at their best.

It was very frustrating for the girls, who'd put all the effort into training, to have to get one of these fellows to ride their horses on the day and to see them fluff things up. So one year the women boycotted race day. Not only did we not go to the races and compete in the Stockman's Race, we also didn't take any of our horses — at all. It was a big disappointment to the blokes. After that demonstration of our displeasure, the programme was somewhat amended and a ladies' race added.

Now things are even more liberated and women can compete in any race. The Stockman's is a fierce, knock-out competition involving a series of heats between two riders, male or female. In the ride-off, the winner of each heat keeps competing until they are the last one left standing. It's full-on, serious, very competitive and one of the highlight events.

Race day at Glenorchy in 2011.

Chapter Fifteen

Rural Vet

A NEW VET IN the district, and a woman! Some of the older farmers on the runs just weren't so sure. Not long after I had arrived at Rees Valley, I came across a chap who needed a couple of his colts gelded. Before vets were available in rural areas, old-timers would just tie a colt up and do the job without an anaesthetic. The pain to the horse would have been considerable. Not only did this fellow not hold with woman vets but he didn't see how I could offer better than that as a method. However, he had a son who thought I could do the job and one day, when his father was away, the son called me up. I went over and did the job using my training and anaesthetics. When the old farmer came home he was just told the job had been done. No explanation was given.

Not long after that episode, I got a call from the old chap to say I was needed urgently. Dave Saxton, the well-known local helicopter pilot, was dispatched to collect me and my vet's kit. Back then, Dave was involved in deer capture and culling, and when I got inside his helicopter it was to find that not only were there no doors, there was also no seat or seatbelt. In addition, the walls and floor were splattered with animal blood.

I sat on the floor, clutched my kit bag tight, and held on for dear life. Fortunately Dave flew level and low all the way. It was my first time in a helicopter and the no-frills ride certainly did not put me off!

In the early 1970s, TB testing of cattle herds was just coming in in this part of the world. It is still TB-free at the Head of the Lake — touch wood. I did quite lot of testing after I arrived, including the very first testing at our back-boundary neighbours, the Branches Station, which is on the other side of the Richardson Mountains from us. They are awfully distant for neighbours — a 320 kilometre round-trip drive via Queenstown

---

Previous pages: Turning in to the driveway to the homestead.

and the Skippers Road — with no through-connection, even for horses, over the range.

When I arrived, I could see that the Branches cattle were 'lively', and I felt a little anxious about working with the herd. However, something totally unexpected came to my aid. The station's cattle yards had been built with poplar saplings, which had taken root and sprouted leaves. The cattle went in willingly, thinking they were getting away in the forest. I was back again three days later to do the testing all over again. The test read: All clear!

A vet's life's not always that exciting, however. Here's a typical day recently. Latte, the Glenorchy Café's cat, is getting old. Although the café has changed hands several times, Latte has stayed on as an important part of the hospitality team. Pale ginger speckled with cream flecks gives him that milky-coffee colouring and perhaps the appearance of being meek and mild. But looks can be deceiving. Only Latte decides which café patrons he will join or give attention to.

Latte has been under my veterinary care for most of his life so I know him well. He — like me, I fear — is getting a few old-age problems. When I'm out on my vet rounds in Glenorchy I try to include a visit to the café to check on him but not to linger for coffee. I usually don't have time. Veterinary calls are fitted around full-on farming activity.

When my telephone rings early or late, I can be pretty sure it will be veterinary enquiries. If the situation is critical, farmers will arrive early in the morning, making sure they catch me before I go out on the station for the day. Their farm vehicles spin up the driveway causing all our working dogs to set up a noisy barking chorus from their kennels. Typically, both my neighbour-client and I are in tattered farm gear. Farm boots are left at the back door. The kitchen table becomes the reception

desk and vet consultancy decisions can happen over the teacups.

Life as a rural vet in the midst of a farming community is quite unlike to that of a city vet practice. I deal with any animals requiring attention, both large and small. James Herriot's book and television series *All Creatures Great and Small* was part autobiography and part fiction, the stories portraying the romance of being a rural vet driving along narrow country lanes and attending to animal problems with any messy business off-camera. Charming! But forget the tweed jacket. Gumboots are the essential kit! Perhaps the only similarities I share with those tales are working in a remote rural district where most clients are neighbours and friends, and the fact that I usually need to go to the animal and deal with a situation out in the field.

In rural practice it's important to limit what you do to the value of the animal. Problems with production animals, such as sheep, cattle or deer, mean counting the cost for the farmer. It is a balancing act of compassion and judgement. And after all these years I always feel a terrible sadness when there is nothing I can do, and an animal dies.

Theoretically I'm on duty 24/7. However, I may be away in the hills mustering so those wanting vet service sometimes have to wait for a few hours before I can attend to their problem. Or I could be on a vet call somewhere at the far end of the practice.

Latte, the Glenorchy Café cat, gets a regular visit. Following pages: Chatting to Anne and Stu Percy during a visit to Routeburn Station to check on an injured horse.

SPEIGHT'S

When I first started Glenorchy Veterinary Services, the needs were for a country vet tending to problems with farm animals. My practice was 90 per cent cattle and sheep, with the remainder farm dogs, horses which were mostly farm hacks, and an occasional cat. The vet who soon after set up in Queenstown specialised in orthopaedic work. What I couldn't fix with a cast — a dog with a broken leg, for example — he'd fix with surgery.

Back then, many properties had no road connection. There was no bridge over the Dart River, for example, so when I did vet work on stations over the river or the lake, I would cross the river on horse, or go across the lake by boat and then get picked up and driven to the station by tractor or truck. My equipment went with me on horseback, by boat, or by tractor — whatever the mode of transport dictated. The most distant calls, as noted, were to the Branches.

Many of the uneconomic small farms in this area have now been sold to lifestyle developers and cut into smaller blocks. With this conversion, the vet practice has grown and changed. Besides production animal work there are many more demands from the increased number of leisure horses and companion animals, usually cats and dogs but also alpacas.

There are now about 200 horses around Glenorchy, including those kept by the two commercial horse-trek companies. Many are kept in paddocks that have been fenced for sheep and cattle. Fencing wire is a hazard, and horses can become entangled in it. They may get frightened by a thunderstorm or a large truck going past and may try to get over or through the fence. Injured horses are now the most common problem for my veterinary practice.

A typical call-out is one I had recently from Routeburn Station, where a four-month-old foal had had a tangle with fence wire and torn its stifle. I donned my gumboots and farm

My clinic is housed in what used to be the single men's quarters on the station.

Left: With a new puppy, Bundle, who is due to have his inoculations. Right: Cynarra Miller's cat receives an injection. Cynarra is the daughter of the late Dave Miller, the visionary owner of Paradise House.

jacket and drove around the Head of the Lake to the station. Both the young foal and its mother had been put in a yard close to the homestead. I leaned over the board fence and watched the pony carefully. Observation showed that some hours after the accident the pony was walking naturally, she was no longer lame, and the wound was showing signs of healing naturally. No intervention was needed.

Dog injuries are another common problem in my practice. Working dogs can fall or get hurt when working out in the steep hills and gullies. The wounded join the kennel row of our own working dogs. We often have an extra guest; recently one was a working dog that had damaged the ligaments of one leg. Putting the leg in a splint enabled the dog to keep walking, ensuring faster healing. I keep the spare kennels for injured and sick dogs that need regular medicine. They are close by and I can keep a watch on the patients and administer drugs as required.

Blake, a much-loved working dog on Routeburn Station, was here for several days and looking sorry for himself with a full elastic bandage preventing him scratching an infected wound on his ear. He took the hospital kennel beside our dog team. He had other dogs close by for company and his owners could drive in and have a chat with him whenever they went past the gate.

It's an extremely varied life as a rural vet. In a single day I may deliver a calf when a cow is having trouble, see a dog that needs an operation, trim the horns off cattle, provide the death certificate for an insurance claim for a working dog that has died from an accident, and then make some house calls. Then there are always the rounds of inoculations for new pets and working dogs.

There is always a need to be calm and aware of how to deal with large animals to ensure safety for both the animal and myself. I stand back to assess a situation rather than rush in.

Being relatively slight and female doesn't make a difference; after all, even the biggest male vet is much smaller than many of the large animals he might be dealing with. It's more about working out how best to assist the animal, whatever its size, working around its temperament and the fact that it may not be used to being handled. I will secure an animal so it cannot do any more damage to itself if it is frightened during treatment. I always talk to the animals and give them a reassuring pat or stroke as I work.

As farming trends change, I have had to up-skill. Even then, things can change relatively quickly. First there was the introduction of deer, which are farmed on local stations mostly as feeder stock for the venison industry. There was the time when many of the stations at the Head of the Lake had stags from which they harvested velvet, but little of this skill is required of the vet practice now. There was also the goat boom, so I had to up-skill on goats, but that's come and gone too.

Extensive knowledge is needed as I am consulting over a very wide range of demands and responsibilities. I may be remote but a phone call to a colleague can often answer my questions. Luckily I've got former classmates who have specialised in specific fields, such as equine. I can go to them for advice if something comes up that I haven't had to deal with over the forty years of my own practice. It's a great network. Lecturers at vet school in Palmerston North are also willing to help with specialised queries.

My diagnostic services are in Dunedin and I send specimens to a laboratory in Mosgiel. Sometimes when I am going to

The ponies in fresh snow. We don't cover them in winter: they grow thick coats and are more than able to cope with the cold.

Dunedin I will deliver specimens or pick up results and talk to the laboratory staff. My reading keeps me up to date on relevant trends, treatments and requirements, but if something unusual crops up experts are only a phone call away.

My charming old building now needs an update; however, it's not in the right position for the next step — a modern clinic away from the station and farm work. These days there is much more veterinary work to fit in on top of the farming. With small farms sold off and cut into smaller blocks, with the Glenorchy population growing and with the sealed road to Queenstown, there is more potential for a full-time vet. My plan is to sell the practice, and it would be better to set up a new clinic further away from the house.

The horses at the gate, waiting for their oats. They are, from front to back, Manu, Miro, Stranraer and Niniao.

Chapter sixteen

Guardians Of The High Country

We Scotts, along with other high-country farmers, see ourselves as the guardians of this special landscape. Living and working in our mountain environment, year after year, has given us a great understanding of this land and all that lives, grows and breeds on it. We have an admiration of and appreciation for its dramatic beauty, and its extraordinary flora and fauna. We also acknowledge the impact of the seasons.

Extensive pastoralism, the high-country farming practice of stocking small numbers of animals who graze over a vast area for short periods of time, is low impact. It is a benign way of managing the land. Conservation and farming are natural partners in the high country. Our belief is that the two work well together. Something I personally have found difficult as a high-country run-holder is that for many years the environmental lobby groups have increasingly portrayed run-holders as a threat to the land. This portrayal suggests that high-country farmers have no awareness of conservation practices and are hell-bent on locking up the land and locking out the public. Farmers at the Head of the Lake and beyond are tired of these confrontations and feel co-operation would bring more understanding.

We have strong personal attitudes about sustainability, animal welfare and the health of the ecosystems on Rees Valley. Our livelihood depends on a healthy environment. We don't just talk — we *do*. We walk and walk for hours and for miles; we recycle — not just the kitchen rubbish but almost anything and everything on the station gets turned around and used again and again; we plant; we nurture; we go out of our way to assist local conservation projects; we drop firewood at the station's huts to make it easier for possum

---

Previous pages: The Rees River, with the Humboldt Mountains at right.

trappers to get out and do their best at trapping. We are not alone in feeling that the black-and-white division between what is deemed to be conservation land and what is regarded as farmland is unnecessary.

I am committed to conservation. The last thing we as a farming family would do would be to endanger the land on which we work every day. Many environmentalists have outdated information driving their stance, which clouds the public's understanding of modern-day farming practices. In the last couple of years, government officials and ministers have recognised that high-country run-holders are good custodians of the land — and just as good as the state in some cases. Priorities vary and a wide range of land management systems will be a better option than the monopoly scenario. Restricted public access, for example, may be better for flora and fauna.

Recent research and careful examination of historical documents has proven that some widely held and frequently repeated beliefs about aspects of high-country farming were not only out of date but also inaccurate. Burning tussock and scrubland has long been incorrectly recorded as a regular farming method. Historically, this was a short-term tactic.

And the burning by early settlers that did occur was not indiscriminate but rather planned. We now know that those early settlers really had no idea of the damage they were doing, even if they only burnt every other year and in specific areas to remove scrub. The first burning of the forests in the Lake Wakatipu area is thought to have been by early Maori, hunting for moa in the sixteenth century. They, likewise, didn't know the damage they were doing. Burning scrubland is limited now we have the benefit of the 150 years of knowledge that underpins long-term management and our present-day farming practices.

The shingly screes that fan down from the tops of some

mountains in the Southern Alps, especially visible in Canterbury, have long been the focus of much finger-pointing at high-country farmers, with the suggestion being made that this was new erosion caused by farming. It was one of the reasons why, in the 1980s, there was a campaign against grazing in the high country. At first it was waged on the grounds of erosion but science didn't back that up except in a few isolated areas, where obviously burning and overgrazing had been taken to extremes by early settlers, sometimes under government edict. In the high country the last thing you want to do is damage your resource because then you can't feed your sheep.

Data and age recording have now scientifically proven the screes to be 500 years old and some even much older. Photography reveals that certain screes have actually shown improved vegetation cover during the last fifty years. Because science did not back up the erosion claim, the focus shifted to making claims that the high country was more valuable for activities other than farming — water conservation, for example. However, water is very pure when it leaves the farmed high country. It is in good order.

There's absolutely no doubt that great damage has been caused by introduced animals — the 'exotic intruders' that have become pests in New Zealand's remote mountains. We have learnt too late the lessons of randomly introducing hundreds of exotic species. Now it is costly to try to keep so many of them under control and to encourage the native bird populations to regenerate to former levels.

Pests including rabbits, feral goats, wild deer, feral cats and possums destroy our remote mountainous lands and indigenous bush. These pests are indiscriminate consumers, eating out some species, degrading the soil, and reducing

---

Me on The Reef Spur in winter, 1980. We'd gone up to snow rake.

birdlife by disturbing nests and devouring eggs. (To these species is now added Canada geese, which have become pests throughout this area. Flocks can concentrate, feeding on valuable pasture.) In complete contrast, farm animals on the high country are controlled. The stock numbers are selected for the good of the land and the number of weeks that the stock is out on the high country is controlled.

We, as a family working in the high country, are only too aware of the significant funds required to manage mountain land, and reduce plant and animal pests, such as feral goats and wilding pines. Luckily we do not have a wilding pine problem on Rees Valley and over the last few years the population of feral goats has been reduced in the Rees and Dart valleys, and Mount Aspiring National Park. However, numbers constantly have to be kept in check as they breed quickly. Wild goats are indiscriminate feeders; they eat anything and everything. We keep a careful watch on feral goat numbers on the range and we shoot goats to keep the population under control.

In winter, when the possum's fur is at its thickest, the possum trappers arrive. It is a good time to trap or poison the animals as we do not have stock out on the high country and there are very few hikers on the trails. We discuss with the trappers exactly where they will set their traps or lay poison. Traps are set one day and then the trapline checked the next. Given the annual trapping programme on the Run, possum numbers are being kept static or even reducing. Most possums are in the beech forest and scrubland, where they can be a nuisance to birds and where they also eat young tree growth.

A family of three generations stays and systematically works the blocks that are free of stock. They possum as a school-holiday family activity, with Grandad, Mum, Dad and young Justin poisoning and collecting possums. Arthurs Creek Hut, where they stay, is chilly but a great experience.

Moving hoggets through Arthurs Creek Paddock to Long Charlie Spur, for their first taste of the high, high country, in 1988.

Graeme loved this photo he took of a tarn in Arthurs Creek Paddock.

We deliver firewood to them but unless the weather turns we leave them alone to get on with the job. The possumers have no contact with the outside world while they are out there, so we need to go to the hut if a bad weather system is coming through, since rain could ruin their poison line or they could be stuck there without food and in bleak conditions for more days than they planned.

I've been a member of the Otago Conservation Board for eight years. The Board works with the issues and concerns of the community, seeking good practice and solutions that work in the Otago area. Members of the Board are nominated, and appointments made by the Minister of Conservation. They form the link between the department and the community. Board members are drawn from a broad cross-section representing all the districts in the Otago Conservancy. Another factor considered in the nomination of Board members is achieving representation across various areas of expertise, such as marine or high country. At the time I was nominated, the Board wanted a representative with high-country experience. They also wanted a delegate from the Queenstown Lakes District, so I fitted two of the requirements, and added a third as the Board sought gender balance.

High-country farmers themselves had felt under-represented, and they wanted someone who could speak for the group, especially given the ongoing tenure review process.

We've just completed one major task, the Mount Aspiring National Park Plan. It took years to get all the considerations addressed. Now we are working on the Conservation Management Strategy (CMS) for Otago. This document is reviewed and updated every ten years. The previous CMS document was a mammoth file. It was too wordy and unwieldy, and the detail was somewhat repetitive. We are working to simplify the next CMS document and also to co-ordinate with

Southland, Canterbury and Westland by providing similar direction for areas adjacent to the Otago region. A more standard formula for CMSs throughout the country will be the result of planning process involved.

Over eight years on the Board I have worked with other members across a wide number of areas and topics. When major discussions are required we organise community seminars. For example, recreational 4WD and motorbike enthusiasts want to use land that has recently gone from private ownership of high-country leasehold into DOC land under the tenure review programme. Recognised tracks have been used by this group and previously they just asked the farmer for access and usually permission was granted. A seminar was held giving us an opportunity to gather information from this group. Then we provided guidance on how to best manage the 4WD tracks that are now part of the conservation estate.

Another initiative was a marine reserves seminar held in Portobello in Dunedin. Otago currently does not have marine reserves, yet there are some globally important marine features and sealife represented in the province, which we believe should be protected properly. Unfortunately reserve status has not been achieved as yet. We got close but not over the line. Marine is not my area of expertise but I fully support these initiatives. Both these seminars are examples of how the Otago Conservation Board formally seeks information from the public.

At other times people talk to us informally, as they know who's on the Board, and if they've got a grizzle they come and chat to us. I've had discussions with the film industry, which wants to make films on conservation estate land. The film

The Rees River's braided course, looking down from Browns Bush and towards the Humboldts, with Mount Bonpland under a cloud.

companies don't want to wait around to get permits if the weather is fine. They just want to go and film. DOC has now developed a better system for processing applications.

Until recently the outcome of most tenure reviews was that much of the tops of high-country land went into the conservation estate. Once this land is under Crown ownership there is a need to actually manage that land. Often the land parcel becomes part of a new conservation park. Public input is necessary to decide what the priorities are for this land. The priority first and foremost is the protection of native biodiversity. To this is added recreation and landscape values. Then there is another tier of considerations such as utilisation by a power company. These are considered according to their merit and such utilisation can happen only if it's not going to seriously affect biodiversity.

Lessees may volunteer for a tenure review; typically they do so in order to acquire freehold title of one part of the property in exchange for surrendering another part. To date, this surrendered portion of the lease has almost always become conservation estate.

If a lessee doesn't like the offer they can walk away from it. As I've noted, we have been through the process and in our case the government walked away. Our preference — and a goal we are working towards — is to place the Rees Valley leasehold land of high conservation value under a QEII National Trust covenant. The agreement is worked out between the Trust and farmer, considering the purpose of the land. Of course the Commissioner of Crown Lands must be in agreement as HM's (the landlord) rep! Under this system the Rees Valley land could still be farmed, and there might be tourism activities. There

Previous pages: In the cool beauty of beech forest, on the Invincible Mine Track.

would be clear guidelines for conservation and the property would not be broken up and sold off to developers.

This arrangement enables landholders to make sure their property is protected in perpetuity against exploitation. The Trust has set criteria, including visiting the areas under covenant every two years. It works hard at consulting with the property owners and monitoring the covenanted land.

We'd be happy to put the entire leasehold property under QEII covenant, with The Run under a management plan drawn up for our grazing regime. This plan would be reviewed every five years. If we wanted to do something different then this would go for consideration to ensure there was no harm to the values the covenant was protecting.

We think that multiple use of the land is an excellent option. Farming income has subsidised land management and conservation, including pest-control programmes, and has allowed recreation. To us, this seems a more sustainable system for New Zealand to consider. The model for operating recreation on land under a QEII covenant or under a leaseholder's tenancy is well tried and in place on DOC land now. Just as there are concessions in national parks for film crews, walks and even bungy-jumps, I see no reason why private landowners shouldn't be allowed to operate the same way, that is, to accept concessionaires that are appropriate on the property and take from them a contribution which goes towards the management of the property. Currently DOC takes a concessionaire's fee. This system works. Our challenge is to work out how to make the conservation and recreation aspects pay their way on Rees Valley Station, instead of being totally subsidised by the sheep and cattle farming.

CHAPTER SEVENTEEN

Our High-country Heritage

High-country farming is the heart and soul of New Zealand, our heritage, and a great tradition, but we are in danger of losing this way of life. Since tenure review there are fewer high-country properties. There used to be about 300 but this number has dropped to around 200. When runs are cut up under tenure review, and the tops go into the conservation estate, then a smaller freehold parcel is left that may be uneconomic. The temptation then is for the remaining land to be sold to developers, who further cut it up.

There's been a long history of the New Zealand government looking at land holdings and deciding they are too big, forcing the leasehold blocks to be divided. When that happens the pieces are again usually uneconomic, unless major changes are made to the land, and the farmers go broke. In the past, neighbouring farmers might have stepped in and taken over the remains of the next-door station. The property then reverts to being a large block. Or the uneconomic piece is made into tiny lifestyle blocks — roast goose instead of golden eggs!

If more and more mountain land goes into the conservation estate, especially into national parks, there are complications. As a nation, we are reducing the areas which had previously been used for productive activities. I'd suggest that the country needs a balance of both private land and public land. These days New Zealand has an outstanding amount of land in the conservation estate, including nearly all of the tops, tussock land, glaciers, and rock and ice areas. Now, with this huge land bank put aside, a different approach is needed as a balance in the remaining high country.

I do agree that the threatened coastal lowland forests and wetlands need protection under the conservation estate. I am

---

Previous pages: A field of newly baled oat straw. Oats have been grown on Rees Valley Station for over a century.

certainly keen to support that. But more mountain land? We could leave a few areas of the high country for recreation that doesn't interfere with nature, the conservation or landscape values.

These activities can cost in terms of investment yet there is little recognition from the government of the value added to the land. We found this out when we developed the Invincible Snowfield in 1995. We made the commitment at a time when people were telling us that high-country farming was in ruins and that we had to get into tourism and adventure activities. So we put effort into the ski field development but found it incurred extra costs in rent and insurance — both required by the government — and that tourism needs huge amounts of capital and time to make it work. Our priority was farming, and we had full-time workloads already.

From this experience, we believe a slightly different approach would work: for example, a portfolio of low-key tourism activity, environmentally friendly and low impact, that somebody else operates as their main focus and business. We'd provide the venue in just the same way that DOC provides the venue for concessionaires on its estate. We wouldn't own the potential tourism business as it makes more sense for us to concentrate on our core business — farming — and a tourism operator can then concentrate on their own business. This would have the potential to contribute to the country's economy and, through the station taking a fee, would also assist with the costs of managing the land.

The current situation is that areas of our land are accessed by the public on their way to and from the Rees-Dart Track. Thousands go across the leasehold land in summer. This access is free, with no contribution being made to the management of the land. Yes, the walk does create work for us: we pull stranded vehicles out of fords, pick up rubbish, have to constantly check on the state of our huts, and more. There are also wonderful walk options on other areas of the property that could be opened up, with concessionaires taking guided groups.

If we had a different tenure arrangement we could develop more recreational facilities such as our environmentally friendly snowfield. There are many activities that have minimal effect on biodiversity. Horse-trekking, for example, one of the key tourism activities around Glenorchy and Rees Valley, has little impact, just the guide 'singing' his song as he works through his patter for visitors and a few hoof prints on the sand and soil.

We are keen to offer private access for walking and study trips for people interested in alpine botany and science. The snowfield hut, which can host groups of eight to ten people, could be used in summer for this purpose. There is so much to see and do near the hut; it is remote and set right in the mountaintops. There is almost nothing similar to it on private land in New Zealand. Individually tailored trips might look at glacial landscapes like the hanging valleys or perhaps the alpine flowers blooming, among them the numerous species of mountain daisy, buttercup, eyebright and gentian. Some buttercup and daisy species are in full flower from November to December, daisies bloom from December to March, with a peak in January, and gentians flower from late January into April — some even in spring.

Helicopters obviously are not popular in places where the public are expecting quiet wilderness hiking and walking, especially in the neighbouring Mount Aspiring National Park. Yet we are aware there are people who have built businesses on being allowed to fly over and into the park for various activities such as glacier landings and heli-skiing. We're following the government's lead by wanting to use our land like this. The

Jim Mora preparing to interview Kate, right, and her friend Helen Ahern for The South Tonight *TV show in 1988.*

other advantage to this arrangement is that it allows for activities that are not welcomed in the national parks areas. And after all, helicopters tread lightly on the hills. There is no need to change the landscape, to build roads or make trails.

I think New Zealand need lots of different baskets of eggs so that if someone trips up then we don't lose the lot in one hit. Government and taxpayer-funded management of land is more difficult during the economic slowdown. DOC has had to prune its management and some key activities. There are projects that DOC has put huge amounts of time and effort into, some of which — including pest control — may not be able to continue because funding has been cut. Who will step up to do these?

There are other ways of doing things rather than always the government paying, or rather, the taxpayer paying. Some of these projects are based on the qualities of Kiwi self-help and determination. A good example is the wilding pine experience. Pine seeds are spread by wind and the wilding trees that subsequently grow can take over a whole area; these wilding trees have become well established throughout the South Island, including in iconic tussock landscapes, and their unchecked spread has become a huge concern. Now there is a joint venture between various groups involving a huge volunteer base and local contributions from people with the passion to remove wilding pines from their own area. This activity will continue because people want to remove these trees to preserve the landscape, without any consideration of payment. Volunteers are keen to keep spending their weekends ripping out wilding pine trees as they are proud of their country and want to do their bit. I remember back to when I was at Massey University. Even then, forty years ago, the students organised weekends dedicated to ripping pine trees out of the scoria on Mount Ruapehu.

It's likely, in the future, that the contribution from government

Diane, Kate (on Morning Star) and me near the homestead in 1980.

will be less and the contribution from volunteers will be more; yet I believe that's a good thing because people like to be involved. In a 'them and us' situation, where we, as a nation, think that since we have paid our taxes then DOC, or some other department, can do the job, we tend to lose interest in or ownership of the responsibility to care for our country. If we recognise that we can make a difference and do something worthwhile, it's much more rewarding. It's our land, and we should be looking after it.

We are committed to upholding conservation values on Rees Valley alongside running our farming enterprise. There are benefits in these two existing together. As an example, well-timed grazing of areas where weeds are an issue can have excellent results. In some areas, once the stock is permanently removed the weeds increase.

Grazing leases used to cover the riverbeds, including the Rees River, but now this grazing licence has been revoked. The change means that the regional and local council and DOC can now spray to manage the lupins and willows, but I think it would be much better to just have a few old cows occasionally wandering along and keeping the willows down rather than putting chemicals into the waterways.

There's a saying, 'it's hard to be green when you're in the red'. And that's very true: conservation costs a lot of money. If high-country farming provided strong returns for our farm products and running costs weren't so huge, then we could do more conservation projects such as replanting and changing shelter belts to native instead of exotic trees.

If you want to keep stock off an area that has been re-planted in natives there has to be a fence. This is why it's a bit complicated doing conservation plantings on farms when the

---

Old gates in the cattleyards.

fencing costs are so high. Although I really like natives, I believe there is an option for both native and exotic plants in particular areas of this district. Taking out all the exotic trees would change the landscape. Increasing numbers of visitors spend time in our district in the autumn because they like the autumn colours; there is a festival in Wanaka and Arrowtown which celebrates precisely that. There is always a balance to take in any decision.

We have our own ongoing native forestry project. I collect seeds from native plants around the property; these are species that survive and are successful here and so it makes sense to continue with them where we can. I propagate my own plants in a shade house, which provides protection for the young plants from winter frosts and snow and summer heat.

We actively contain weeds and pests, taking a low-impact approach. For example, matagouri, the native thorny plant that used to be called wild Irishman, has been considered a pest but I believe it is useful if left to grow. In settler days it was burnt off — a very big mistake, as some early explorers and farmers found out. When it regrows, matagouri comes through low down and is very prickly. It tore at the legs of drovers and shepherds, as well as at the legs of their horses, dogs and stock. Early diaries tell of the tribulations of trying to get through regenerating matagouri. The land then became unusable as there were too many thorns for stock. It became a liability. When you fight matagouri, it fights back!

If you leave matagouri, as we do, it eventually grows upwards and develops a trunk. The branchy part is above the height of grazing animals and can be useful for shelter. When deep snow falls, the ground under the matagouri is sheltered;

McDougalls Creek, running down from McDougalls Spur to join the Rees near the Invincible snowfield helipad.

as the thaw starts, drips from the branches hasten the clearing of the snow around the trees, giving stock access to grasses. Insects for birds and the bats flourish in the scrubby country — another good reason to leave well alone.

We have drawn a line on the property, designating where it is suitable to have exotics. On the flat land, The Farm, there are shelter belts and ornamental trees. It is a mixed landscape with some wonderful exotics as well as native plants. The northerly area of the run is gorse- and broom-free, and it's our priority to try to keep this area free of those weeds. The area of Rees Valley nearest to the national park, the bulk of the property, is where we make considerable effort to control anything that might be a threat to the park. We work hard to remove weeds, such as old man's beard and the willow trees in the creek beds.

We would like to do even more in pest control than we already do but these programmes are costly, so we are always juggling what can be afforded this year or next — always with a long-term view.

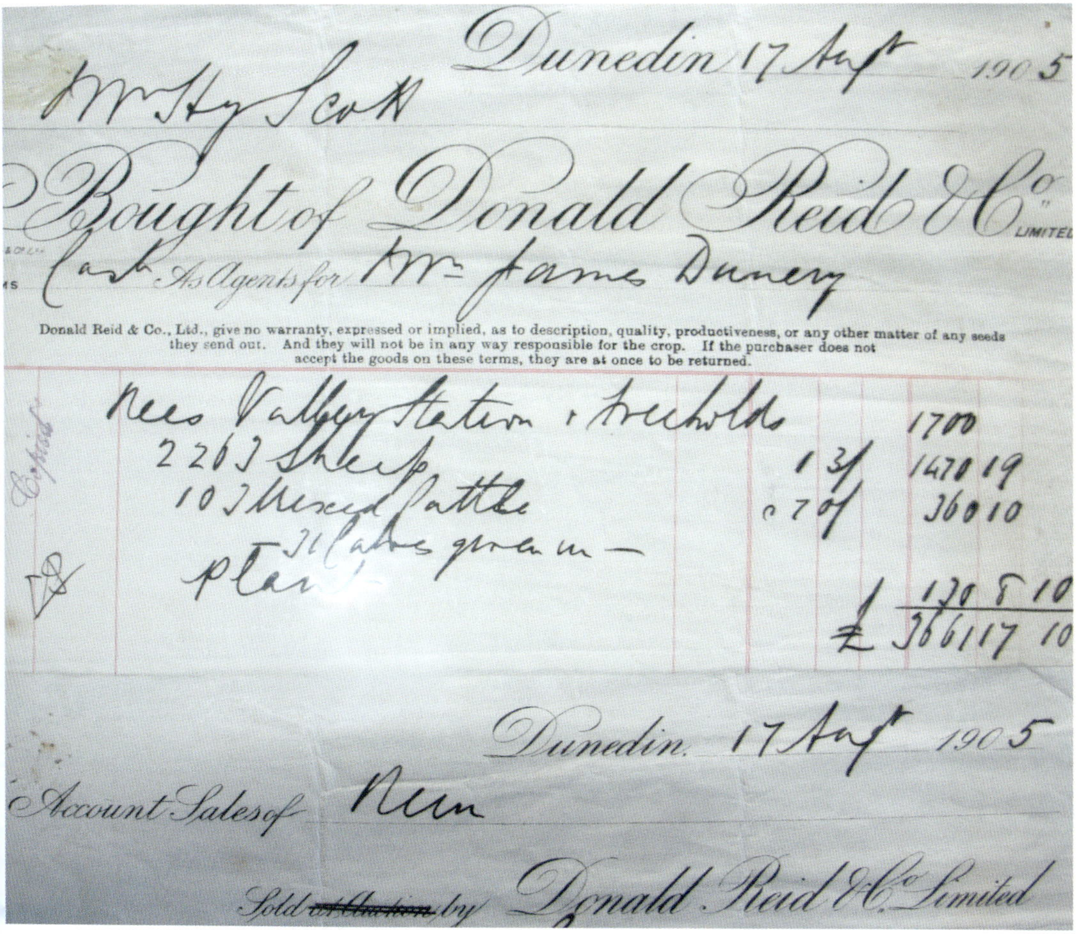
Dunedin 17 Augt 1905

Mr Hy Scott

Bought of Donald Reid & Co. Limited

Cash As Agents for Mr James Dunery

Donald Reid & Co., Ltd., give no warranty, expressed or implied, as to description, quality, productiveness, or any other matter of any seeds they send out. And they will not be in any way responsible for the crop. If the purchaser does not accept the goods on these terms, they are at once to be returned.

Rees Valley Station & Freeholds		1700
2263 Sheep	13/	1470 19
103 Mixed Cattle	70/	360 10
31 Calves given in — plant		
	£	130 8 10
	£	3661 17 10

Dunedin 17 Augt 1905

Account Sales of Rees

Sold by Donald Reid & Co. Limited

The bill of sale for Henry Scott's purchase of Rees Valley Station's leasehold, freehold and stock in 1905.

Chapter eighteen

More Than A Farm

It wasn't long after I arrived at Rees Valley all those years ago that I came to appreciate that the station is home to so much more than cattle and sheep and we few humans. I am reminded of this every spring when I am out on the quad bike and I see my first pied oystercatcher of the season. For Head of the Lake locals, the oystercatchers' arrival heralds spring. The worst of winter is over. The birds come here to nest and breed, and later they will fly back to the coastal estuaries and harbours.

I notice how the oystercatchers often segregate themselves in a field, away from the plovers and paradise ducks, and how they tend to choose the middle, where the sheep have just been; we suspect this is where there are extra grubs and insects. They nest in the shingle and silt-banks of the Rees, but also, at times, right in the middle of our paddocks! Eric has to be extra careful driving the tractor around them. A potential threat is cattle, horses and sheep trampling their eggs or even their chicks, but they seem to fend them off. Just as they will fly up and squawk to drive people away from their nesting sites, so they will with farm animals. And they always seem to prevail — the stock give them a wide berth.

Or my attention will be caught by the call of that striking and noisy bird, the New Zealand falcon, the karearea. We delight in the many New Zealand falcons that can be seen around the station, especially on the farm area next to the river. There are more New Zealand falcons per kilometre in Glenorchy than almost anywhere else in the country. They can fly at incredible speeds, over 100 kilometres per hour. And they can take mice, other birds, even roosters. It's important to keep a roof on the hen run!

Falcons use the power poles along the Rees Valley Road

Previous pages: The Glenorchy lagoon near the Glenorchy township.

The karearea, or native falcon, an impressive bird of prey, is a constant presence in the Rees Valley.

and also our tall exotic trees as eyries from which to spy on us. They'll dive, and you'll just see a small puff of feathers as they take a whole bird. Alas, sometimes that means oystercatcher chicks, too, if the parent is not nearby. And there's one bit of my beat on Broad Spur paddock where falcons have been nesting in a rocky outcrop for season after season, and where more than once a falcon has had a go at me. The sound of them swooping down right close to your face is more than a little frightening, so I tend to walk through that pinch in the rocks at a fast clip.

One winter I was walking along the riverbank when I disturbed a falcon that had been sitting in the long grass. I'm not sure who was the more surprised, the bird or me. It flew up right in front of me with a huge beating of its wings. Startled though I was, I had time to observe its beautiful plumage, the variegated markings of brown and tan, with the wings undershot with a brilliant orange. It was a joy to watch as it soared across the river and then disappeared into the beech forest.

Rees Valley Station's forested areas support a wide variety of fauna, including many native birds and insects, and its habitat is a feeding zone for an endangered native bat, the pekapeka.

I love being in our bush. I love how you can be one minute in the tussock, under a bright sun and clear blue sky and then, with only a couple of strides, step into a cool quiet world. The deep green, silvery and olive-green leaves of the beech trees filter the sunlight; a walk here is a walk through dappled shadows. It feels timeless. And it's affirming to know that our pest-control work on the property is helping to maintain and build bird numbers.

Tui song rings out as they command their territory. Yellow-breasted tomtits, robins and fantails flutter around like nosy-parkers checking on the 'foreigner' in their territory. They will come close to chase the grubs and insects disturbed by our boots in the leaf litter. Bellbirds are common, too, and can often be seen arguing with the tui over nectar; they're especially fond of the native fuchsias.

At night I hear the ruru or morepork calling from the bush. By day, closer to home, tauhou or silvereyes are regular visitors to the garden. Usually they arrive in small groups, so several dart to and fro in the same bush. Once the garden's berries are gone they will move on through the valley to whatever else is ready to harvest. We notice blackbirds outside the kitchen window, especially when the holly hedge is in berry in mid winter.

Low on the station's hillsides and in the creek valleys, the indigenous forest is a mix of several varieties of beech: red, silver and, predominantly, mountain beech. Research suggests that much of the eastern side of the Southern Alps was likely to have been in forest such as this after the thawing of the last ice age, but at some time in history, whether naturally or deliberately lit, massive fires burnt these forests off. It's thought that Maori also used fire as a means of hunting to drive moa out of the forest hundreds of years ago. When early Pakeha settlers arrived most of this forest had long gone. Rees Valley Station fortunately still has patches of indigenous forest that survived the fires.

The forest and scrub insects attract rare native bats, which fly in to feed at twilight. Pekapeka, a long-tailed bat, is found only in New Zealand so it is a very special little creature. It is one of only two native land mammals and is about the size of a mouse.

We've happily been able to assist pekapeka research. In the early years of the research programme, volunteers and rangers used bat-detection 'smart boxes' which picked up the bats' location calls, the ultrasonic sounds that help them navigate and forage while flying at twilight and night. The smart boxes tracked the movements of the bats, and through these movements there was an effort to estimate the population. To do this monitoring, pairs of us would walk set routes along the roads into the back country twice a year.

Above and below: Tiny pekapeka bats with transmitters fitted. Opposite: At the Glenorchy DOC office with staff member Barry Lawrence, who was well known throughout the Lakes District and Otago for his visionary conservation work, including on bats. Sadly, Barry died in November 2011, aged only 60.

Threatened
small leaved tree daisies
of Wakatipu
Olearia species

Recently DOC staff have been able to attach transmitters to some female bats and track them back to their roosts. Numbering the roosts and learning about the population in each one has brought researchers closer to determining the population more accurately. What has been intriguing to discover is that they roost in the Dart Valley and at twilight fly into the Rees Valley to feed — at twelve kilometres, a considerable distance for such a tiny creature.

Not too long ago, two rats, Harry and George, occupied a cage in one of the station's outhouses close to the homestead. These domesticated rats were bought from a pet shop and were quite used to the homely, even privileged, life in a cage, where food arrives every day and a warm nest of torn-up newspaper is frequently refreshed!

Harry and George had an important part to play in research into predation of wild rats, which are such huge pests in our native forests. As part of a project I would place Harry and George, safe in their cage, at various locations around the station. They remained there while infrared cameras were trained on the cage over twenty-four hours to check what animals might try to prey on them. Over two years of the project, alas, no evidence of rat predators was seen.

New Zealand now has several rat-free zones where well-recognised conservation efforts have established a pest-free environment. These include world-leading bird sanctuaries on islands off the coast, where an extraordinary effort is made to keep the islands rat-free and where, as a result of this mammoth effort, the recovery of birdlife, especially the establishment of native and endangered species, has been outstanding. Happily we also have some of those islands at the

The 'A Team', Ling and Rogue, leap from the Land Rover and set to work.

BJH315

Head of the Lake: Pigeon and Pig islands are alive with birdsong as they, too, have become rat-free sanctuaries for birds.

I don't know what Harry and George thought of their research experience on Rees Valley; in any event they were ultimately comfortably and honourably retired from their research duties. Then one morning I went out to discover that Harry and George had been killed in their supposedly safe cage by an enterprising stoat.

Despite our so-called remote location, university researchers from around the world also cross our doorstep and spend time at the kitchen table. Two university researchers from Aberystwyth University in Wales have been visiting over the last few years, studying the Rees River, its flow and movement. They are known locally as the Welsh Wizards.

The homestead at Rees Valley is a practical two-storeyed house with multiple bedrooms. Into these many bedrooms pile family, friends and an ever-changing group of WWOOFers.

WWOOFer stands for 'World Wide Opportunities on Organic Farms', and its volunteers, usually international travellers, are happy to work for free in return for board and keep. Where one can stay as a WWOOFer is usually spread by word of mouth. I remember the day Irish Paddy, who was from a small-holding on the west of Ireland and who helped with the cattle muster, arrived to find the huge Fisher wood burner that we call Grandma Bear going in the kitchen. 'It's just like home,' he smiled.

Paddy had a lot of experience with stock and was a great help to us. Others have had no farming experience at all. Sometimes they're young travellers who've just walked one of the nearby Great Walks and, after coming down the valley and feeling its beauty, decide they want to stay to experience the area for longer. They just knock at the door to see if there is anything they can do to help. These travellers get a bigger taste of the high country than the tourists who just drive past in a rental car and then have a latte in Glenorchy.

Added to the household, then, is a multitude of nationalities that stay for a few days or much longer. For these travellers it's a taste of real New Zealand life and a real New Zealand family as well as an in-depth glimpse into the heart of the high country, which adds so much more to their memories of this country.

I can tell whether WWOOFers are going to be useful or not in about half an hour. The first test I give them is gathering up pine cones to feed the wood stove. It seems simple but it pretty well determines how useful they might be; sent off under the pine trees with a few sacks to fill, little do they realise it's a career test for their future work assignments on Rees Valley Station. Some think they've gathered enough after half an hour; others work away for a few hours and pass the test easily.

However, I always find something for even the most impractical to do! But the results can vary hugely. We once had a couple of Thai girls who completely misunderstood our instructions and sent away the chap who arrived to buy the bull. We also had two British girls who showered twice a day and had the heater and lights on all the time. The electricity bill I received after their visit was quite a souvenir of their time here; we decided after these two not to have WWOOFers in winter again!

The Best WWOOFer Award went recently to a crazy and charming chap from Germany, who became indispensable because of his work ethic and his humour. An academic, with a master's in philosophy and about to start on his PhD, he made exceptionally witty jokes in near-perfect English and kept us in stitches. He became firm friends with Fred, the 'dancing pet lamb' — a particularly appealing orphan we'd been bottle-feeding who would prance about the place on his hind legs.

Many of the WWOOFers stay in touch; postcards arrive and emails trickle into the inbox from all around the world. And passers-through are frequently lucky to encounter the locals. On the drive back from a vet call-out recently, for example, I picked up a Taiwanese hiker who had just walked over the Routeburn Track but had missed the connecting bus back to Glenorchy. He heaved his pack onto the back of the Land Rover and I gave him a lift as far as the station gate, from where it was likely he'd get another ride the rest of the way. So the veterinary call-out which didn't require attention turned out to be a bonus for an international traveller. Perhaps he will remember the brief encounter with one of New Zealand's friendly farmers he met on his travels.

Dropping off a Taiwanese hitchhiker who's pleased to have got a ride closer to Glenorchy.

CHAPTER NINETEEN

Working Dogs

LAST WEEK, KATE BROUGHT eight-week-old Queen home from Temple Peak Station. She's adorable. It was time Kate got a new huntaway in her team; call it succession planning. Queen's a good-looking little dog and currently a roly-poly bundle of fun. We admired the way she settled into her kennel on the first night and didn't cry a bit, even though it was her first time away from her siblings, near dogs she'd never met before and on a new farm with new people.

Both Kate and I have our own working-dog teams. We have more huntaways than heading dogs as that's a better balance for the sort of sheep work we do here. There is more of a need of workers to shift stock in the yards and paddocks and along the roads, and huntaways come into their own in that environment.

My current team is Ling, Maid, Rogue, Jerry and Arthur. Kate's is Tess, Queen, Bonny and Storm. The teams are a demographic mix: youngsters, the middle-aged, the slowing-down and the almost retired. Dogs are like people, with their own traits and characteristics. When pups are young I can pick their personality. I try to work dogs to their strengths, not their weaknesses. And just like people, sometimes dogs don't get on with their masters. Just as there are people you cannot work with, there are dogs that you just can't work with.

But when you have a good dog, then every day you are reminded of how little we could do if we didn't have them. Dogs are lean, keen; like high-performance athletes, the marathon runners of the canine world. Alert and ready, they watch from their kennels for the door to open and their chance to work. Once asked to go, they run with streamlined precision and, if required, leap high over fences and gates. Effortlessly. They also

---

Previous pages: Maid is keen to get out of the back of the Land Rover and off to work.

Kate's new huntaway puppy, Queen.

slink, stealthily, low and slow. I never fail to be impressed.

When I first came to Rees Valley there were heaps of dogs. Doug and Graeme had their own teams and there were several other farmhands and shepherds working here with teams of up to six dogs, on average, as well. The last one to leave left me an old dog to learn with. You start with one dog, learning how to command it and move small mobs around. (This can be useful, especially if everyone else is out mustering or moving mobs on other blocks; a young bloke can step in with his dog to move a few sheep around on the farm paddocks if required. It means that this work doesn't get left to the end of a long day when the others come home from the hills.)

Later I got a pup I called Chip. It is really important to use a short name, usually of one syllable. And when adding a new dog to a team you have to think carefully about their name so that it is quite different from any of the other names currently in use. Otherwise total confusion would reign when commands are given.

Once I got Chip trained I could go out and be useful with the sheep. Gradually I built up a team, and have had my own ever since. Most people can use a dog whistle but it's something I have never mastered, so I use voice and hand commands. This can have its disadvantages. A dog may set off to act on a hand command when I only wanted to get my handkerchief out of my pocket!

There have been many memorable dogs on Rees Valley. For me, the standout dog was the remarkable Sioux. His mother, Lass, was given to me by the legendary deer hunter Wattie Cameron of Makarora, one of the pioneers of high-country deer culling by helicopter in the 1960s. Our daughter Diane named the litter of Lass's pups after American Indian tribes; Apache and Sioux were added to my team.

The first time Sioux impressed me was when I was at the

head of Invincible Creek. I knew I could not get to where I could see the sheep on the far side of the basin and return in daylight, so I asked Sioux to go out on the blind. I was sending him out a long distance to bring in sheep that he couldn't see when he set off. Normally dogs will bring the sheep straight back to their master, but the straight line between me and Sioux was extremely hazardous and steep at the head of Invincible Creek.

Sioux followed tracks until he found the sheep and then carefully guided them back, allowing them to travel at their own pace without pressure, working them around the head of the basin and avoiding the steep rocky creek bed. It was half an hour or so of work. Much of the time he couldn't see me and I couldn't see him. It was an outstanding long run.

Following pages: The dogs on full alert in their kennels, ready for action. From left, Rogue, Maid, Storm, Ling, Bonny and Tip.

Chapter twenty

Recreation In The Mountains

Past the gate of Rees Valley homestead go many of the world's keenest trampers, hikers and alpine walkers. Mostly they are not looking up our driveway but rather focused on the walk that they are about to do, or the multi-day hike they are about to make. The lucky ones get a lift as I move between the Farm and the Run.

These folks are going to and from the start or finish of the Rees-Dart Track and sometimes coming down from the Routeburn or Caples tracks; if they are walking along the road they'll get a lift as far as the gate, where they are highly likely to pick up another lift into Glenorchy or straight through to Queenstown. Up in the Rees Valley the road is open to the public to the Muddy Creek carpark, and beyond that it is a paper road. When the road blocks are stocked we change the signs on the gates to ensure the public knows whether it is to be shut or can be left open.

Via the road to the trailhead, the public can access Mount Aspiring National Park. This has created some extra tasks for us: we are always on the alert for any litter and when the creek level rises trampers will occasionally get their cars stuck in the deeper fords. I often record the number plates of cars parked in the car park beyond the last ford or driving on the road. If hikers forget to sign out after having done the Rees-Dart we can let search and rescue teams know whether the car is gone; it's likely they have driven away in their car and all is well. Awareness of who is in the valley is something to note. With such demanding and rapid changes of weather, it's a good idea to keep a record of who might be around just in case of problems.

Previous pages: A hiker on the Rees-Dart Track, near Arthurs Creek. Thousands of hikers trek across our property every year. This page: Arthurs Creek and the two-stand woolshed at 'the hub of the universe'.

The four- or five-day Rees-Dart Track is considered one of the South Island's finest tramping experiences. The route follows the braids of the Rees all the way through the valley and it attracts thousands annually; the numbers rose once the Glenorchy–Queenstown Road was sealed. But they are often in for a shock. One of the hikers we picked up last autumn when we were out herding the Herefords was 'French Simon'. He'd been five days on his own walking from the Dart Valley, out across the tops and then down into the Rees Valley. It was already into autumn and he had been waist-deep in snow on the tops for quite some days and had got very wet. When we came across him, he'd spent those five days in wet clothing and had met no one else on the track. He was delighted to get down to Rees Valley and then find that the run-holder was there. He got a lift on the back tray of the Land Rover to the main road, a slow trip as the cows and calves were being mustered.

Recently a fresh-faced Czech couple setting out to do the hike from Rees Valley, up across the tops and down into the Dart, got a lift from me all the way to the parking area near the start of the track. It was autumn, and already snow was covering the high areas. They would also find that they would be tracking through snow for much of the way.

On the station itself there are several walks, including to the DOC reserve (land given by the station to the Otago Goldfields Park in 1970) at the Invincible Mine. (Mountaineering on the property is not in huge demand as many find Mount Ferguson, the tallest peak, is not that exciting or challenging.) The half-day walk to the Invincible Mine zigzags up the hill through beech forest to emerge at the

---

Looking north towards the upper Rees flats. The Hunter Stream runs into the valley from the left at the head of the flats and to the right is Mount Clarke towards the northern end of the Forbes Mountains.

old mine site, which was in its heyday of operation in the late 1880s. A few relics remain at the site, including the Berdan pans used for breaking down the quartz and remains of a water wheel. Graeme remembered many more items being about when he was a child. However, with so many visitors and no guardians on site, a few people are tempted to take an object as a memento.

Having so many people on the station is not without its issues. We understand that the area offers a world-renowned alpine and back-country experience. Most people keep to the tracks, which are marked with orange marker poles. The issue is that now with sealed roads and big 4WDs the area is available to so many more people — thousands pass over our station in the summer season on their way to the national park and the two tracks. Many do not understand where they are going or the difficulties they may encounter, and they do not understand that they are walking over station land, and the need to keep away from stock. Some new DOC signs, recently erected, are a great help to us. There are those who are just not prepared and those who are there for unlawful reasons. More freedom campers than ever are travelling in vans without facilities. They can leave rubbish and 'deposits' that no one wants to find in this pristine environment.

We are happy to have public access but need to know where people are going. Foot access is usually granted if people take the trouble to contact us (or indeed the landowner of any station) about the land they want or need to cross. We understand how to the public it might appear that the land belongs to no one, yet it is private; without fences it isn't clear that the Rees-Dart traverses station land where animals can be disturbed. Our animals are out on the pasture for weeks, and know that when we arrive they will be expected to move and be mustered. Sometimes the presence of humans gives the animals the idea that mustering is about to start and they get confused and may move to where they think they need to go. We later find they have moved off their pasture and can be in an area without good nourishment.

In early times this area was remote and only accessible by a few. But now with the increased population of Queenstown and so many travellers in the region that has all changed. Many more people have 4WDs — fun to drive, but we hope that drivers would consider the farmer's investment in construction of tracks and maintenance of the land and seek our advice before heading out. A vehicle bogged in wet weather leaves its mark, sometimes for years.

Trail bikes have opened up the back country for many young people. At their best they are okay; at their worst they bring noise pollution and ecological damage to fragile landscapes. There are just a few who ruin it for the majority by frightening stock, leaving garbage and lighting fires where they shouldn't, and by scarring the landscape or hooning around the riverbeds, threatening ground-nesting birds such as wrybills, dotterels, terns and oystercatchers.

Many visitors do not understand the unwritten creeds that previous generations have grown up with. Those who regularly work, live and play in the high country develop that understanding over many years; the lore and ethics get passed down and passed around. Newcomers arrive with no history or understanding; they come from countries where 'remote' tends to mean that a café, a shelter and a bathroom are not far away. I don't believe some of them understand how isolated our mountains are or how challenging the trails can be, even though the DOC information, if they have checked it, will be explicit about the level of fitness and skills that are required.

Following pages: Remains of a trampers' shelter built in the ruins of the Invincible post office.

Chapter twenty-one

A Hat-tip To Mr Rees

I WONDER WHAT WILLIAM Gilbert Rees would think of Queenstown, Rees Valley and the Head of the Lake now — 150 years on from when he first walked into it?

William Gilbert Rees, Wakatipu explorer, pioneer, earliest settler and entrepreneur, was the run-holder who first leased land for pastoralism at Wakatipu. The eastern side of the lake, including Rees Valley, all became part of the massive acreage of Rees's Run. He brought the first merino here. He built the first buildings beside the lake on the site that was to become New Zealand's alpine tourist resort town, Queenstown.

As an entrepreneur, Rees might approve of Queenstown now. He was quick to see a business opportunity. He would be amazed at the visitors and travellers from all over the world, the entrepreneurs, the food and wine, and the excitement and buzz. Just 150 years. It's an astonishing achievement in such a short time. Rees, too, would be celebrating!

If I took him out on a drive along the side of the lake it's likely he would acknowledge again its extraordinary beauty and that of the mountains, and his vision for the land. He would admire the easy journey to the Head of the Lake compared with his four-day struggle on a crudely made raft after which he then bush-bashed through almost impenetrable scrublands and over steep mountain slopes. Glenorchy, the village, would be somewhat familiar to English-born Rees: little cottages, rustic gates, gardens and trees, and people on horseback.

He'd come to Rees Valley Station, The Farm, and he'd see the heritage of English farming: neat, flat, vivid-green fields, huge old trees and hedges, bountiful fruit and nut trees, bales of hay, sheep, Hereford cows, calves and bulls, horses and dogs. He'd find The Run untouched, remote and stunningly beautiful. He'd

---

Previous pages: The Leap Ridge reflected in a puddle near Lennox Falls and the track to Kea Basin.

A self-portrait of William Gilbert Rees, at age 26. His vast land holdings earned him the moniker 'King Wakatip'.

once again see the valley of tussock grasses, rivers and creeks and waterfalls, the hillsides of beech forest, hundreds, perhaps thousands, of years old.

He's a fascinating character to me. William Gilbert Rees grew up in the English countryside, yet his wasn't a farming family. He hardly knew his naval-captain father, who was away during Rees's youth aboard peacetime and wartime vessels. After his private tuition, followed by boarding-school education, Rees took a teaching post at a boys' school — one of the few employment options available to a young educated man without property. Rees, however, sought more. Like so many other young men of his generation, he decided to seek his fortune on the other side of the world, in the new colonies of Australia and New Zealand.

In March 1852, Rees set sail for Australia to try his luck, first managing a farm property. But it soon became apparent that he would need to move to the next frontier: New Zealand. Squatter farmers had already settled most of the suitable farming land in Australia.

Once he had arrived in the south of the South Island in 1859, Rees wasted no time in putting together an expedition to the interior of the unmapped southern regions that lay inland from the coastal settlements of Oamaru and Dunedin. Rees set out with a group on horseback, with pack horses carrying provisions, travelling from Dunedin to Moeraki, and then turned inland in January 1860. His expedition at first headed for Wanaka, thinking that district offered opportunity. They accepted hospitality at farms along the way. In Wanaka, friends of Rees had already obtained leases and had begun farming. After considerable adventures and near-perilous attempts to make their way any further south, many of the party turned back.

But strong-minded and resolute, Rees and his friend Nicholas von Tunzelmann continued on through the Cardrona Valley and eventually up and over the Crown Range. The journey took days and days of effort, including several failed attempts to find the pass that would take them through the range to the other side, where lay terra incognita, unknown land. Imagine their excitement when they saw Lake Wakatipu and the land around its shores, and realised how suitable it was for pastoralism.

The area had been inhabited from time to time over the previous few hundred years by nomadic Maori who travelled the routes over the Alps to collect greenstone on the West Coast. It's likely Maori also visited the area hunting for moa. And Rees and von Tunzelmann, in fact, were not the first Europeans to visit Lake Wakatipu; a few, albeit less than a handful, had approached from the south, travelling with Maori guides. However, the region was yet to be settled and was not mapped. There was, for example, an awareness of the lake's existence but not of its true shape or impressive size.

Rees and von Tunzelmann camped at the small bay where the Queenstown Gardens are now. They wanted to travel further up the lake but it seemed impossible to journey alongside it: the steep hills were thick with undergrowth, high tussock and scrub, and in places the hills plunged straight down to the lake edge.

So the two friends, amazingly, created a raft from driftwood and bundles of flax-flower stalks. The lake was so whipped up by storms that it took them three days of attempts before they were able to set out in this frail craft from Queenstown Bay. Once they got to the Head of the Lake, Rees was able to see more land that he thought might be suitable for pasture — land that is now Rees Valley Station.

One of the last things Rees did before departing the Head of the Lake for the return journey to Dunedin was set alight the scrubby matagouri. Whole hillsides caught on fire, and there was so much smoke that Rees and von Tunzelmann had difficulty finding their way out of the area.

The journey back was another arduous multi-day adventure which they almost didn't survive. They arrived back in the then tiny settlement of Dunedin, emaciated, gaunt, exhausted and weather-beaten. Rees wrote to the local newspaper about the land they had seen and then he and von Tunzelmann quickly put their names forward for the leases of the land around the shores of Lake Wakatipu. Government policy was to see land settled and developed — the economy of the new colony depended on increasing productivity and growing sheep numbers for wool.

Nicholas von Tunzelmann gained the rights to property on the western side of the lake; Mount Nicholas and Mount Nicholas Station are named after him. Rees and his partners, the Australian graziers George Gammie and Colonel William Lewis Grant, obtained several leases, totalling about 240,000 acres, on the eastern side of the lake, including the lease for land at the Head of the Lake.

Rees eventually settled near the middle of the lake, at what is now Queenstown. He purchased a whaleboat near Invercargill and had it transported overland to the lake by bullock wagon. Rees then used this boat to send supplies up and down the lake to his stockmen.

The boat became invaluable. The miners, who came after the gold find in the Arrow River, would surely have starved if Rees hadn't already been established and farming at Queenstown Bay, where his homestead was called The Camp. He had systems in place to deliver food supplies, including meat and potatoes. But ironically it was also the goldrush that deprived Rees of the lease of land where he had built his homestead.

Diamond Lake, between Rees Valley and Paradise.

It was chosen as the site for the new town of Queenstown and he was forced to leave only eight years later, after having done so much for the area, including starting a timber mill and establishing buildings for his home-station headquarters.

Rees's name now is embedded in local lore, both at the Head of the Lake and in Queenstown itself. He is credited with founding Queenstown, and Frankton is named after his wife Frances. (Rees named the eponymous river after himself; and von Tunzelmann, the scion of a Baltic German family, named the Dart after an area close to his own home in England.) Rees became known locally as 'King Wakatip', a sort of lord over the local kingdom of Wakatipu. His vast acreage and leases were called the 'promised land of Rees'.

To establish North Run, Rees had hired Alfred Duncan, for whom Mount Alfred is named, as a stockman to take the 3000 merinos he had had transported by ship from Australia to his North Run station at Wakatipu. Duncan was a new arrival who had just endured, as he wrote in his own history of the area, a voyage from England that he described as '107 days of ocean surroundings. On board a death rate of 36'. Thank goodness he later sat down to write his history, *The Wakatipians,* as it vividly depicts the extreme hardship early shepherds and drovers endured.

Duncan set off with thirteen horses and one mule, driving the sheep through river crossings, along unmarked tracks and over mountains. He wrote:

This was my first experience of travelling with stock and it was of sufficiently rough nature to give me a good idea of what lay before me in the life I had chosen. Leaving the Waitaki we brought the sheep past where Oamaru now stands — at that time, the city consisted of a shepherd's hut only — when we reached the Otepopo river we struck off the road, and heading south west prepared to cross the Kakanui range, Gilbert having conceived the idea that he knew of a short cut that way which would enable us to reach Coal creek in a marvellously short space of time.

We were provided with a piece of salt beef, and tea and sugar, and did not lay in a supply of bread.

All day we toiled on over rough and rocky ground drenched to the skin with heavy rain and unable to see any distance owing to the thick mist which had settled over the ranges.
At last when it was quite dark, we managed to put the sheep over a stream and on to a spur
of the hill beyond.

When he was near what is now Frankton and Queenstown, he discovered that the speargrass was so sharp that his trousers and legs, and the limbs of his horses, were cut to pieces. The rocky ground cut the dogs' feet badly.

He then had to take the sheep around the hills beside the lake to North Run, another five days. The drive was an incredible feat. And once he got to the Head of the Lake he went straight to work, as this excerpt from his diary shows:

We moved our camp further up the Lake as we thought a better site for building a hut was on the spit opposite the mouth of the Rees River where a large lagoon opened off the Lake and we built our first establishment that was built at the head of the Wakatipu. It was 12 feet long by 10 feet wide built of sod on the spot thatched with grass and with a chimney at one gable which was about four feet high. I was left with a month's rations of flour, tea and sugar.

Just surviving appeared daunting. Not for nothing did he coin the expression, the 'toilsome nature' of the country.

Rats were of plague proportions and made life a misery for the men attempting to set up station life. Their supplies had to be stored in metal tins and the state of their meagre provisions of flour and tea were also in doubt, having at times been 'off' before they were delivered overland by bullock team from Southland to Kingston, at the far end of the lake. These supplies had already had a journey of several days. Next stop was to load them onto a boat to go up the lake.

Run-holders who arrived slightly later with wives and children made use of abandoned goldminers' shanties and even forsaken and neglected hotels before building their own houses. A settler history, *Sparkling Waters of Wakatipua* by Florence Mackenzie — one of many that we have in the homestead — describes these well:

The first homesteads were of wattle and daub, and these were very good sound homes. The pioneers cut down trees, choosing the number required as nearly as possible the same thickness and length. These were placed perpendicularly against the framework side by side. A mixture of clay and chopped up tussock was thrown or plastered on this. When it dried it was white washed, and the process repeated once or twice a year, as required or as time permitted. Often it was the wife who did the white washing, not only of the old fashioned fireplaces and ceilings, but also the outside of the home. After those primitive homes came the stone houses and very fine buildings they were. The thatched roof under our modern one is a work of art and in its time withstood many a severe winter snow storm.

It was all for the wool. The name of nearby Blanket Bay is thought to have come from the first shearing done on North

Run, next to the lake. The shearers spread blankets on the ground so the wool clipped off the sheep would not get entangled with the leaves, twigs and grass nearby. A woolshed was later built for North Run where Glenorchy is now.

There was a constant round of reapplying for leases and when, in 1872, the oddly termed Waste Lands Board was abolished and the department of Lands and Survey established, all run-holders then had to reapply again. Rees was followed by Butement on run Number 346, the North Run lease. Then, as I have described, there were two more leaseholders, Valpy and Dunery, before the arrival of Henry Scott in 1905. Butement had built up extraordinary flock numbers on North Run, which included what are now Earnslaw and Walter Peak stations; they reached 42,000 before an exploding rabbit population wreaked its havoc, declining to 17,000 in 1866. At the time, government edict was to fuel stock numbers to fuel the economy. There was little consideration of how much stock the land could sustain. They simply had no idea; it was virgin territory.

Then, under the Liberal government of Richard Seddon, came a change in policy. The vast South Island runs were to be divided up so more small farmers could occupy land. At the Head of the Lake this meant that Butement paid more than the going rate to acquire some blocks as freehold; that is how the area around the homestead, The Farm, at Rees Valley Station became freehold land.

Much of the area where the township of Glenorchy is based was surveyed out into 50-acre sections, which people could freehold. The idea was that a family could live on 50 acres, have a cow and a few sheep and the man of the house could work in the nearby gold and scheelite mines, in the timber industry or as a farmhand or shearer. They could earn an income and feed themselves from their piece of land. But in the end, only some of the 50-acre blocks were taken up; mining declined as did timber milling, and it soon dawned on people that in an area with such a tough climate and generally poor soil quality, 50 acres was never going to be an economic unit. Many of the blocks never sold and were later thrown back into the adjacent pastoral leases. Glenorchy, however, survived.

Kinloch, across the lake from Glenorchy, photographed some time in the 1880s.

Chapter twenty-two

The Head Of The Lake

Established in 1862, Glenorchy village has just reached the milestone of 150 years of European history. That history includes that of early explorers and adventurers and fortune-seeking goldminers. It was even, at one time, the home town of Clark McConachy, who in his day was the world's top billiard player, and who has a creek named after him. McConachy was the son of Glenorchy's storekeeper and billiard saloon proprietor. He left school at fourteen to work in the billiard saloon and by age seventeen he was making breaks that had him noticed. In 1951 he took on the world as a professional and succeeded, becoming the world champ five times.

Glen Orchy in Scotland is a glen, or valley, around the River Orchy. This tranquil little valley, surrounded by alpine peaks, is considered an unmistakably serene and beautiful part of the Scottish Highlands and is famous for climbing, walking and salmon fishing. It's not difficult to see why the New Zealand connection to this namesake was made.

And it's all Scots, with one 't', round here. Early Scottish settlers named many of the creeks and streams around the district, using the Scottish term 'burn', as in Buckler Burn and Route Burn. Campbelltown is the name of the tiny settlement of only five houses just outside Glenorchy, and across the lake is the township of Kinloch. Then there's Mount Earnslaw and Ben Lomond. You'd almost think a man in a kilt might appear any minute!

Glenorchy is located among some of the most spectacular alpine landscape in New Zealand. In the opinion of early explorers and settlers, the scene here is equal to, if not better than, other highly acknowledged landscapes around the

---

Previous pages: The old freight shed on the Glenorchy wharf, which dates back to the pre-road days when the lake steamers were the only freight and passenger link with Queenstown, Kingston and the world beyond.

The Glenorchy Library, established in 1892 by community-minded locals.

world. It's right on the edge of the lake. Close by is a huge fan of shingly river braids and a bird-filled lagoon. Above all this beauty soars an amphitheatre of snow-tipped mountain peaks. It's stunningly beautiful, and the camera doesn't lie. Every click is a masterpiece.

It's been called a postage-stamp-sized town. And as one of the most popular meeting points for visitors and locals alike is the café, which was once the post office, the correlation appears easy. The postage reference also works with the stamp-shaped town plan: trim streets in a neat grid form a small square, edged at the lakeside with curvy shingly bays.

The census puts the population of Glenorchy at about 430 residents but the locals don't believe it's this high. Hazarding a guess, we reckon it's probably about 350 people.

Horses and riders are sometimes as frequent as motor traffic and are the very good reason for the wide grass verges and the lack of kerbing around the town.

The town's handful of heritage buildings remind us all of local history. Both new and old are in company with comfortable house-buses, artfully recycled huts and corrugated-iron sheds converted to homes.

Remains of the North Run Station woolshed and cookhouse were in existence for many years. Local legend suggests that, like many structures, much of the material from these two well-established buildings was recycled and reused. The cookhouse became the Glenorchy Store. The roof and timbers of the woolshed are thought to have been recycled for other nearby station buildings and the stones from the walls for wharf supports.

However, many other small and fascinating remnants of early times remain, including the picturesque and important Wharf shed, the school building, and the Mission Hall church. Then there is the miniature old Glenorchy Library, which local

Glenorchy hotels in the 1880s.

Miners at the entrance to the Invincible mine in 1882.

residents believe was once one of the smallest public libraries anywhere in the world! The early station owners tended to be well educated and sought to provide learning and books for their families, and a social life through church rather than hotel bar. Books were keenly shared and a small lending library began; Henry Scott was on the organising committee in the early 1900s.

Glenorchy has welcomed waves of newcomers seeking prosperity, but many of them have encountered disappointment; it's been tough here over the years. The area's population has flowed, then ebbed, then flowed again. All too fast, the town erupted when goldminers, scheelite-miners, sawmillers and shepherds arrived in a few short years in the late 1800s.

Glenorchy, and nearby Campbelltown, were boom towns for a few short years. Prospectors searched the rivers and creeks for gold, and later two waves of miners worked the hills for scheelite. The miners were tough men who liked a drink; local lore has it that in 1888 every occupied building in the town was a licensed hotel!

It's not just because we have the remains of a mine on our station that I find the early mining history of the Head of the Lake interesting. Gold was mined out of the creeks to the point of exhaustion and later scheelite extraction became uneconomic, although this metal got a second and third run during the Second World War and the Korean War, respectively. Scheelite is the ore essential for making tungsten, which was used for making large gun barrels, being strong and heat resistant.

In the mine's heyday, the stamping battery could be heard going day and night up on Mount Judah, behind the town. It wasn't an easy way to make money. The miners worked in gruelling conditions, in mountainous terrain, where it was

---

The ruins of the scheelite battery on Mount Judah — undergoing restoration at present.

often better to be working inside the tunnels than spending off-work hours in a bleak, wind-blown shanty or hut. Fog came around the peaks, the area was prone to flooding, and snow could arrive unexpectedly, delivering drifts of up to 6 metres. No wonder the miners took to the tunnels and kept busy.

Some of the mines were very high in the hills. Miners made their way up first by pack horse and later with just a pick and shovel, and later still with a wheelbarrow. Sheer determination and some ingenious solutions allowed the mines to be worked with supplies carried in by a pack horse pulling a sled, plus a hefty dose of pure brute strength. The carefully mined materials were recovered from the mountain by a huge flying-fox contraption, an engineering masterpiece.

Families were reliant on the mine, with fathers and sons working on the hill. One day a week, Sunday, the women and children would visit their husbands, fathers and brothers. Through this hardship was forged a camaraderie which is remembered in personal tales.

It pays to remember all this when you take a walk up there today. It's an easy climb up the side of Mount Judah, through the Whakaari Conservation Area. The track takes you past the scheelite battery (processing plant) and the old workings. Beyond are the remains of the Bonnie Jean Mine, to which the huge flying fox was linked.

Glenorchy has seen out the tough times of the boom and bust of goldmining, depression and despair in the 1890s, the rabbit plagues, and then the Depression of the 1930s. Sometimes these catastrophic events forced residents to leave without anything to show for their extraordinary efforts. Over the last 150 years, Head of the Lake locals (some of whom have been here for over five generations) have needed a fair dose of toughness, resilience and self-sufficiency. It's pretty civilised

The main street of Glenorchy in the late 1880s.

The paddlesteamer Antrim *at the Glenorchy wharf in May 1905, having ferried 6000 sheep from Glenorchy to Mount Nicholas Station over five days.*

these days — and nowhere near as isolated — with a sealed road, electricity and a café but, even so, you need to be a little bit self-reliant when there's a power cut or a flood — floods occur more frequently than big snows.

Adaptability is another essential trait if you are to survive in Glenorchy. Many locals juggle several careers, both to make a living and to fulfil community needs. Duayne Insley, the deputy head of the volunteer fire crew, once turned up at the homestead to ask what size boot one of our WWOOFers took. When he found that the boots he had with him fitted the WWOOFer, he told him he was now part of the Glenorchy Volunteer Fire Service, part of the team. In a small community locals find themselves in all sorts of roles. Someone may be the boss of a tourist operation and also be first call for the First Response ambulance team. They might also do some cattle mustering on their days off.

Glenorchy has always been a Scots place, and local life is still strongly laced with Scottish character and heritage. The hardy immigrant settlers from Scotland brought a diligence and strong work ethic with them, an attitude that has remained. These days we don't work half as hard as those early Scottish farm workers and miners, but even so, when times are tough it sometimes feels as if we are Scottish crofters on the hills!

This work ethic and resilience looks, straight-laced at first glance but behind this façade, richly colourful individuals shine. When Irish Paddy says of the crowd at the café bar, 'I met some characters', it says something! He would know. Colourful characters often choose to live in places out of the mainstream. They certainly bring creativity and individuality to this small community.

And it wasn't all work and no play back in the old days. New Year's Eve is one of the major annual festivals celebrated by the Scots. They brought this celebration to New Zealand, and for a time First Footing (when neighbours and friends visit one another after midnight to bring New Year good wishes) set the Head of the Lake district alive very, very early on the first morning of the New Year. The station's farm diaries tell of the first footers arriving at 4.30am! After the first footers had visited the homesteads and homes along the road, they would all end up at the Earnslaw Station homestead for breakfast. It got to be too huge, with too many people coming from Queenstown to take part, and therefore had to be ended.

These days, after hours, the locals enjoy any excuse for a celebration. In true Scottish spirit an occasion is never let pass if it could offer some fun. Community festivities for families and children include pet day at the school, fishing competitions and flower shows, and lots of fundraising for essential services.

Locals tend to be unplugged (although I suspect that fewer and fewer of even the older generation are technophobic old peasants like me!). We mix hard work with fishing, kayaking, horse-trekking, hunting and more. Some locals are hard-out guides. Every car seems to be a practical 4WD splattered in mud. Outside the library, the school door and the café there's likely to be a pile of gumboots and farm boots — all worn and muddy. Outdoors adventures, sometimes extreme, are daily living and breathing, not just a once-in-a-lifetime activity on a holiday junket.

We all love the local radio station. Mainstream radio is not heard in these parts; there is no reception. Glenorchy Country radio station 89.2 is a little bit country and very local. Driving in to Glenorchy from Queenstown, you can pick up the signal

---

Following pages: Glenorchy in the late 1880s.

GLENORCHY, HEAD OF LAKE WAKATIPU. Nº 226. MORRIS

1864. UPPER REES VALLEY - LAKE WAKATIPU.
MUIR & MOODIE. DUNEDIN, N.Z.

from Bennetts Bluff. The DJ Chas Drader, a long-time radio man, operates from his home-based studio and often asks Glenorchy residents what we want to listen to, scheduling our choices of music regularly. In two hours of talking every weekday, Chas will cover the road report, relying in winter on the reports of the grit-truck driver, who is on the road early in frosty weather. He also reports on the weather by looking out his window! Actually, he does have a weather station, and therefore has the temperature right here in Glenorchy.

My Land Rover radio is always tuned to 89.2. I enjoy the local ads and varied music, and Chas slips in a few Dylan favourites just for me. He knows his audience and knows that many residents are often working outside in trucks and farm vehicles, where a bit of music livens up their day.

Music has always been a part of life round here; almost always, music for entertainment has been provided by anyone who thought they could make music. Miners and sawmillers turned their hands to banjos and the ubiquitous saw blade — dragging a bow across the back edge of the saw to emit notes. We have one of the old hand-made banjos, which is now a hundred years, or more, old. Dances were held in the town's garage, the only place with a floor space big enough for the band and dancefloor. There were dances every week, usually on a Friday night. In the 1930s, the dance band was built around the Scott family. Jean played the organ at church and the piano at the dances, along with cousins and uncles and aunts from Temple Peak Station who played on a variety of instruments. Later on, Graeme, who played slide guitar, joined Jean as one of the band's musicians.

Everyone joined in because everyone had to. After all, until

Previous pages: This photograph of the Upper Rees Valley in 1864, only four years after Rees claimed the vast North Run, shows the mouth of Twenty Five Mile Creek flowing into the Rees further north than it does today.

The Glenorchy band in the 1930s. Jean is in the middle of the front row.

the road was sealed it was an incredibly remote place. The locals had to make their own fun.

Now, every day, travellers arrive to spend a couple of hours in the village. Some will go down to the wharf and look at the lake, or have a coffee at the café or a beer in the pub. Some will stop at the First World War monument and note how many young men from the Head of the Lake gave their lives in the Great War. Others are here for day-long or multi-day adrenaline adventures: jet-boating, horse-trekking and mountaineering with local operators and guides. It's a busy wee place at the end of the road.

The Glenorchy Post Office and First World War memorial in the 1930s.

Chapter twenty-three

Slice Of Heaven

GLENORCHY TOWNSHIP HOLDS THE key to adventures in the extraordinarily beautiful alpine surroundings. But this is nothing new. The Head of the Lake was there right at the beginning of tourism in New Zealand. As early as the late 1860s tourists made their way here, for walking, fishing and sightseeing.

First came landscape painters to record scenes in oil and watercolour, then later cameras and film captured the splendour. Richard Bryant opened a guesthouse and hotel at Kinloch, just across the lake from Glenorchy, in 1868. The name Bryant dominated local tourism as the business expanded into transporting and guiding visitors, first by horse and dray and later by bus, from Kinloch on into the Routeburn Valley. The Glenorchy–Paradise route was run by the Knowles family, and later the Smiths, who ran the garage in Glenorchy.

Steamers on the lake gave a romance to the trip. An essential lifeline for the stations, the steamers also opened up travel on the lake for tourists. The boats called at all the stations around the lake, which had their own wharves, adding to the fascination. In 1885 three hotels were built all in the one year in Glenorchy to cater to the influx of tourists.

In the old days, horse-and-trap and riding parties came out to see Lennox Falls, which are on the boundary between Earnslaw Station and the Mount Aspiring National Park on the western side of Rees Valley. Named after English peer the Right Honourable Walter Charles Gordon-Lennox, the spectacular waterfall drops like a ribbon of water off Mount Earnslaw in two long descents. Here the valley narrows to provide a majestic scene of jagged peaks on three sides. An occasional hawk glides high, then swoops for prey. Mist sprays up and over the deep green of the beech forest fringing the falls. It would

Previous pages: The Upper Rees Valley showing Mount Head and Mount Clarke and the valley of the Hunter Stream.

Arcadia was built in 1907 as a tourist house. Tourists would come up the lake by steamer and then be conveyed there by horse and trap, and later by bus, for afternoon tea.

have been an adventure. You can imagine how wonderful it would be to be part of a riding party here.

In the 1900s the centre of the tourism activities shifted to Glenorchy, en route to Arcadia and Paradise, a 14-mile trip. Paradise was the major tourist attraction for early holidaymakers, before the Queenstown–Glenorchy road opened. Visitors would sail up the lake on a paddle steamer and disembark at the Glenorchy wharf, to be collected by buggy and dray.

The guest house at Paradise was established in 1891, a change from its original use as a rather glamorous and salubrious 'crib' for the Mason family (William Mason was New Zealand's first government architect, and later the mayor of Dunedin, and built Paradise House in 1883). The hospitable Paradise House and the elegant, two-storeyed Arcadia nearby — an impressive large wooden house with turrets built between 1906 and 1907 for Joseph Cyprian Fenn as a tourist house — became sought-after destinations for well-heeled tourists seeking alpine splendour who didn't mind the two-hour buggy trip to get there. They were also a favourite port of call for local residents, who visited for tea, to dine and to play cards. Today, continuing in the tradition of alpine hospitality, one of New Zealand's finest lodges, Blanket Bay, is located on the lake edge just out of Glenorchy.

Bryant's continued as the tourism company well into the next century, changing from horse and dray to specially modified open-topped buses. There was always a scramble to be one of the lucky ones to get on this popular roofless bus to see the magnificent scenery. Even the buses themselves had originally arrived by steamer. Like all other vehicles, they were loaded on the steamer after being driven onto the wharf at the Queenstown end and then lifted aboard by the boat's steam winch. At the Glenorchy wharf, the hoist lifted them from the top of the steamer, ready to drive off the wharf and away. The unsealed road, different from the current route, took the long

Open-topped buses took tourists to Paradise House.

Run by a trust, Paradise House still operates as a guest house, and has just had a fantastic restoration that fulfills the vision of its former owner, the late Dave Miller. This early photo shows how true to the original the restoration has been. Scenes for The Lord of the Rings *and* The Hobbit *were filmed not far from here.*

way over an older bridge spanning the Rees River further up the Rees Valley. The bus trip would end at Paradise House, run by the Aitkins and then the Veints, for tea or a three-course lunch.

These tours were a thriving sightseeing activity that continued until the road opened to Queenstown, after which Paradise went back to being a secret hideaway. Now a few adventurous tourists set off to explore the area in their rental cars or motorhomes. All visitors are likely to be astonished by the scenery, but Glenorchy residents giggle at the frequent comments: 'I've driven to Paradise but there was nothing there.' Exactly. That is why it is Paradise, surely?

Why Paradise? A few ideas are put forward. Some say it may be because of the plentiful paradise shelducks in the area. But the favoured reason is that the name reflects the surroundings. There is even a small area nearby called the Garden of Eden, a clearing in the bush in which you can stand surrounded by the mountains.

An observant visitor will note that the names of peaks around the Dart Valley and Paradise area, named by early explorer and surveyor James McKergow, are entirely classical in origin: there's Nomus (Greek God of Laughter), Erebus (God of Darkness), Somnus (God of Sleep), Nereus (God of the Sea), Nox (Latin for night), Amphion (the Musician), Minos (King Minos of Crete), Pluto (God of Infernal Regions), Chaos (Space), Sarpedon (son of Jupiter and Europa), Niobe (daughter of Tantalus and Queen of Thebes), Cosmos (the Universe), Tantalus (son of Zeus and King of Phrygia), and finally Cerberus (the three-headed dog guarding the Underworld).

There is even a Mills & Boon romance featuring this special location! *A Place Called Paradise,* by New Zealand romance writer Essie Summers, was published in 1967. Essie came for the

Advertising to tourists in the late 1930s.

TRAVEL IN COMFORT AND SEE
THE BEAUTIFUL SCENERY OF
LAKE WAKATIPU
TWIN SCREW STEAMER "EARNSLAW" (ALL WEATHER VESSEL)
RUNS TO HEAD OF LAKE ON MONDAY, WEDNESDAY & FRIDAY.
LEAVES QUEENSTOWN – 8.30 A.M.
ARRIVES QUEENSTOWN – 6.30 P.M.
RETURN FARE
17/6
ONE OF THE FINEST ONE DAY STEAMER TRIPS IN THE WORLD.
CONNECTS WITH MOTOR SERVICES TO ROUTEBURN VALLEY & PARADISE.
Interesting Commentary given over the Ships' Public Address System.
MORNING AND AFTERNOON TEA AVAILABLE ON THE STEAMER.

holidays and then wrote a romance tale set in the area, which, in her telling anyway, was a giddy social whirl of dances and parties for the heroine, Annabel Lee.

Annabel Lee had been brought up in an orphanage, but had always longed to be able to trace her father, and at last she decided to follow the few slender clues she had to her real identity. They led her to the most beautiful place in the world — where Annabel not only found what she was seeking, but met Gideon Darroch, the most wonderful man in the world, too!

Paradise and the nearby beech forest was used as the location of Parth Galen and Lothlorien when Peter Jackson was filming *The Lord of the Rings* trilogy. When he was still at Dunstan High School in Alexandra, Eric was selected to play an orc in *Lord of the Rings*. The scenes he was in were filmed in Poolburn, in the Ida Valley, and he remembers the experience as 'lots of waiting around and then fighting scenes', which might be just a bit laconic for the hundreds of diehard fans who, somewhat amazingly, keep coming to the Head of the Lake every year to see the locations for themselves. Paradise House has recently been restored and you can still book a room there. In fact, in all its long history it has never *not* been an accommodation house. That's quite a record!

I arrived at the Head of the Lake after the era of the steamers. The construction of the road had ended their long reign. The tourist trip on the TSS *Earnslaw* is now only a faint suggestion of the activity up and down the lake of years past. The steamers, the first of which arrived in 1863, were the lifeline for all who lived around the lake. Wood-fired, and later coal-fired, they were the link to Queenstown, Frankton, Kingston and beyond for local merchants, run-holders, miners and sawmillers. Wagons drawn

Tourist cars lined up outside the Glenorchy Post Office in the 1930s, waiting to collect tourists and take them out sightseeing.

A mob of sheep is transported up the lake by TSS Earnslaw *in about 1960.*

by horse and bullock, and, from 1878, by train, carried people and goods from Dunedin and Invercargill to Kingston. From here, goods went on to Queenstown and the lake settlements, including to Glenorchy, by boat. Once tourism was thriving, the service aboard the private steamers didn't meet the standards of comfort expected by tourists so the government bought the local steamer company and placed steamer services under the control of New Zealand Railways.

The arrival on the lake of the TSS *Earnslaw* in 1912 created quite a stir. The whole vessel was built in Dunedin and then carried in pieces to the lake before being assembled. Her beautiful interior of varnished timber, with velvet-covered seating, added to the style and splendour. Such was the excitement about her maiden voyage that a holiday was declared all around the lake, so the locals could celebrate.

The Scotts were regular passengers, and relied on the boat to carry stock (dogs and drovers would drive the sheep up the wharf straight onto the steamer), bales of wool, cars, trucks, tractors, and friends coming to visit or for weddings in the family. Doug and Jean sailed away on the TSS *Earnslaw* for their honeymoon, waving to the well-wishers on the wharf from the roof of the boat. On board were many of the guests, who had travelled up by the steamer prior to the wedding. Motor cars also arrived this way, including the first of the family's Fords, and each car was driven onto the wharf at the Queenstown end, lifted on board by a hoist, then lifted off at Glenorchy to drive down the wharf there. Photographs of these momentous occasions appear in the family photograph albums.

The TSS Earnslaw *on the lake in 1951. She was first launched in 1912; a public holiday was declared all around the lake for her maiden voyage.*

Loading a car and freight onto the Earnslaw *in the 1930s. Following pages: The boardwalk across the Glenorchy lagoon wetland.*

Chapter twenty-four

Looking Ahead

THIS MORNING I DROVE out to collect a young university student who is on the station doing work experience, to bring her back to the house for lunch. I'd had her clearing blackberry vine from a fence and then burning it. She's been doing a science intermediate year at Otago University and has applied for veterinary school at Massey. Just like me all those years ago she's waiting for her results, to see if she's been accepted. I can imagine the nervous expectation she feels.

Things are so different at Massey now: the majority of the students are female. But the passion for animals and for working with them hasn't changed. I can see that by the way she plays with Queen, the huntaway puppy, when she's been fetched from her kennel and brought over to the kitchen for a bowl of milk.

After lunch I take the Land Rover across The Farm to move a mob of hoggets. I take Rogue with me, and he moves the sheep off down towards the gate to the next paddock on his own while I check a long ditch to make sure no sheep have fallen in and been left behind. I find one lying on the ditch's bank, unable to move; its left foreleg is broken.

Rogue has things under control ahead of me so I lift the sheep onto the Land Rover tray. I'll take it back to the yards near the house, where I've got a few sick beasts I'm nursing back to health, and put its leg in a splint. The bone will knit together and it will be fine to walk on again before long.

A motherless calf needs feeding, so I mix up his milk and carry it out to him. He sucks on it as if there's no tomorrow.

Kate comes back from school at the end of the day. There's a bit of a buzz in the village, she reports. Peter Jackson has

---

Previous pages: The Humboldt Range in winter, viewed from the entrance to Whakaari Conservation Area. I look at these mountains from both my bedroom and the kitchen windows.

Eric feeding his pet calves in about 1988.

arrived at the Head of the Lake and has driven up to Paradise, where there are now about 400 crew assembled for *The Hobbit* filming. Cars have been going up the road all day. We locals have become a bit blasé about the fact that our neck of the woods has become so famous worldwide through *The Lord of the Rings* and the dozens of television commercials and advertisements shot here. It's a nice little distraction that provides plenty for us to talk about and speculate on.

Meantime, I'm planning the week's work. Eric hasn't been able to come home from Makarora for the weekend and we will miss his skills. Kate is going to have to drive the tractor and do the spraying. She's been calling him to find out exactly what she needs to do. I'll be up the valley tomorrow to check out the lambs. Farming goes on.

During the time that I have run this station, the South Island high country has been through so many twists and turns, has been buffeted by so many storms — political, economic and natural. The 2011 election was pivotal. Say what you like about National, this government has been kinder to high-country farmers. We felt that Helen Clark's administration was actively hostile. Station land was being transferred to conservation estate at a great rate and there was no sympathy for our issues or understanding that we are capable, responsible custodians of the land. Or that we are part of a long farming heritage that is important to New Zealand's identity.

It's a heritage that I am proud my family is part of. But farming the way we do is changing, and there are fewer long-term owner-operator families working in the traditional way in the high country. There were over 300 high-country stations, including Rees Valley, under Crown pastoral lease when the 1998 Crown Pastoral Land Act came in. Since then that number

Drenching a ewe at the woolshed yards.

has decreased by more than one-third. Some stations have gone through tenure review; some are halfway along the process. Other stations with good water access have gone to dairying. Some stations have gone to overseas owners and local managers run the properties.

The high country runs on long timeframes. Accountants think in terms of five years and short timeframes. We are different; all our decisions are made for long-term sustainability. And because we are deeply linked and committed to this place. Kate, Diane and Eric also have that special bond with Rees Valley, just as Heather in Germany still does; although she has spent so much of her life away, there is still a deep attachment.

I am in my sixties now and some days I keenly feel the hard work. I know that there will be other changes. More new technology. More new ideas. This is what keeps the buzz going and gives us new things to learn. At Rees Valley, once there were only horses and men (and a few women!) in the hills on foot, then came 4WDs, and now we have helicopters. But one thing that won't change is the hard work required of us as farmers.

When I first came here I could have days off because we had staff. Back then a station worker worked five-and-a-half days, with time off on Saturday to play football. On Sunday they were expected to go to church and have a respectful, restful day. All that changed as staff went, and in the financially difficult times the work has been up to us to do. Next year Kate is coming to work alongside me full time, and I am looking forward to having more of her help. The women are still hard at work on Rees Valley Station!

When I drive past Bennetts Bluff on my way home from a Conservation Board meeting elsewhere in Otago, or from a trip into Queenstown to do the food shopping, there will be a couple of cars stopped and tourists will be out marvelling at the view and taking photos. I've seen it now thousands of times, but I still slow to admire the view up the lake. Those big mountains still thrill me.

We share this land with many others — the trampers and the climbers, the riders, the jet-boaters, the hunters and the scientists. For all the time of the Scotts' occupation, all these recreational interests have been allowed and managed. Conservation, as it has been understood at different times, has also been actively supported. Weeds and pests and fire management and huts cost money, subsidised by farming. The question for the future is how farmers can be allowed to develop some commercial recreational options on their runs to generate additional income so that these other, public-good activities that benefit all New Zealanders can continue. We are reluctant, or unable, to invest while we are subject to political change.

We are lucky to see this beautiful part of New Zealand every day, in all its different seasons, and to know it intimately. It inspires us, it challenges us. The Head of the Lake and the Rees Valley is our home. I like to think there will be many more generations of high-country farmers here.

Following pages: Red beech forest on the way to Paradise; Horse trekkers ride alongside the Rees River.

Acknowledgements

I would like to thank Geraldine for the work she has done researching topics I mentioned in our discussions. She has done a great job of putting a bit of Rees Valley into print. I would like to dedicate the result to all those people and animals who have made the station what it is today, and hope future generations will not judge harshly our methods of producing food and fibre sustainably. Fashions in farming, conservation and recreation change; let us always look for best solutions, not convenient lines on paper.

Iris Scott

There can be no better place to live and work than amongst the alpine majesty of the Glenorchy area. Writing and researching this book opened a window into the extraordinary day-to-day life of high-country farming. Thank you to Iris for welcoming a writer into the midst of her family and her work, and a special thank you also to both Eric and Kate; my time with you all was a privilege, and laced with unforgettable memories.

Thank you, also, to Nicola Legat at Random House for offering me the chance to learn so much more about the heart and soul of New Zealand's farming heritage.

Geraldine O'Sullivan Beere

Image Credits

Aaron McLean: 21, 86, 122, 341.
Alexander Turnbull Library: 291, 311 (bottom).
Amanda Hasselman: 28.
Dick Veitch: 262 (top).
DOC: 262 (bottom).
Geographx: 24–25, 26–27.
Geraldine O'Sullivan Beere: 1, 4–5, 8–9, 12, 15, 16–17, 29, 31, 39, 40–41, 47, 52–53, 65, 68–69, 72, 81, 82–83, 89, 91, 92–93, 97, 101, 102–103, 107, 111, 113, 114–115, 116 (top and bottom), 118–119, 125, 127, 133, 135, 136–137, 140, 143, 146 (top and bottom), 148–149, 154, 156–157, 158–159, 161, 170–171, 174–175, 180, 194–195, 198, 203, 204–205, 210–211, 214, 216–217, 219, 220 (top and bottom), 223, 225, 226–227, 236, 238–239, 242–243, 250, 253, 255, 256–257, 259, 265, 269, 270–271, 276–277, 278–279, 281, 283, 286–287, 288–289, 295, 302–303, 305, 309, 334–335, 336–337, 344, 346, 350–351, 349, 352.
Gilbert van Reenan: 182–183, 342.
High Country Horses: 209.
Lakes District Museum: 307 (top), 311 (top), 314–315, 321, 325, 327 (top and bottom), 331 (top and bottom), 333.
Mike Langford: 2–3, 6–7, 33 (bottom), 191, 322–323, 346–347.
Miles Holden: 177 (top and bottom).
Real Journeys: 329.
Scott family: 33 (top), 35, 37, 43, 45 (top and bottom), 49, 56, 59, 60–61, 66–67, 74, 76, 128–129, 139 (top and bottom), 162–163, 166–167, 185, 188–189, 200, 207, 230, 233, 234, 247, 249, 273, 319, 339.
Museum of New Zealand Te Papa Tongarewa: 301, 307 (bottom), 316–317.

Bibliography

Miners in the Clouds: A hundred years of scheelite mining in Glenorchy, Julia Bradshaw, Lakes District Museum, 1997.

Head of Lake Wakatipu, School Centennial 1894–1984, Centennial Committee, 1984.

Land of the Mountain and the Flood: A contribution to the history of runs and runholders of the Wakatipu district, Peter Chandler, Queenstown & District Historical Society, 1998.

Birdlife of the Mount Aspiring National Park, Peter Child, National Park Series Number 4, Department of Lands & Surveys, 1981.

The Wakatipians; Early days in New Zealand, Alfred H. Duncan, 1888. Reprint, Lakes District Centennial Museum, 1969.

Exotic Intruders: The introduction of plants and animals into New Zealand, Joan Druett, Heinemann Publishers, 1983.

The Story of Wakatipu, D. A. Knudson, Whitcombe & Tombs, 1968.

Central Rocks: A guide to the geology and landscapes of Central Otago, Daphne Lee & Jane Forsyth, Geoscience Society of New Zealand, 2008.

No Idle Rich: The wealthy in Otago & Canterbury, Jim McAloon, Otago University Press, 2002.

Whose High Country: A history of the South Island high country of New Zealand, Roberta McIntyre, Penguin, 2008.

The Sparkling Waters of Whakatipua: The story of Lake Wakatipu, Florence Mackenzie, A. H. & A. W. Reed, 1948.

Making Sheep Country: Mt Peel Station and the transformation of the tussock lands, Robert Peden, Auckland University Press, 2011.

Spirit of the High Country: The search for wise land use, The South Island High Country Committee of Federated Farmers, 1992.

Queenstown: A geological guide, I. M. Turnbull & P.J. Forsyth, Geological Society of New Zealand, 1988.

Historic Sheep Stations of the South Island, Colin Wheeler, A. H. & A. W. Reed, 1968.

Tikumu, the Maori name for mountain daisy.
She's the spitting image of her father, Matai, the stallion from Arcadia Station who fathered all our little black ponies.
Following pages: Looking down on the head of Rees Valley from the Invincible Mine ruins.

A RANDOM HOUSE BOOK published by Random House New Zealand
18 Poland Road, Glenfield, Auckland, New Zealand

For more information about our titles go to www.randomhouse.co.nz

A catalogue record for this book is available from the National Library of New Zealand

Random House New Zealand is part of the Random House Group
New York London Sydney Auckland Delhi Johannesburg

First published 2012

ISBN 978 1 86979 820 8

Text and cover design: Katy Yiakmis
Front cover image: Aaron McLean
Cover images: Geraldine O'Sullivan Beere
Printed in China by Everbest Printing Co Ltd

This title is also available as an ebook